Ergebnisse der Mathematik und ihrer Grenzgebiete

Band 65

William Browder

Surgery on Simply-Connected Manifolds

Springer-Verlag New York Heidelberg Berlin
1972

William Browder
Princeton University, Princeton, N.J. 08540, USA

AMS Subject Classifications (1970):

Primary 57 D 02, 57 B 10, 57 D 10, 57 D 55, 57 D 60, 57 D 65
Secondary 57 D 40, 57 D 50, 57 E 15, 57 E 25, 57 E 30, 57 D 90, 57 A 99, 57 C 99

ISBN 0-387-05629-7 Springer-Verlag New York Heidelberg Berlin
ISBN 3-540-05629-7 Springer-Verlag Berlin Heidelberg New York

 Library of Congress Catalog Card Number 70-175907. Printed in Germany. Printing and binding: Brühlsche Universitätsdruckerei, Gießen.

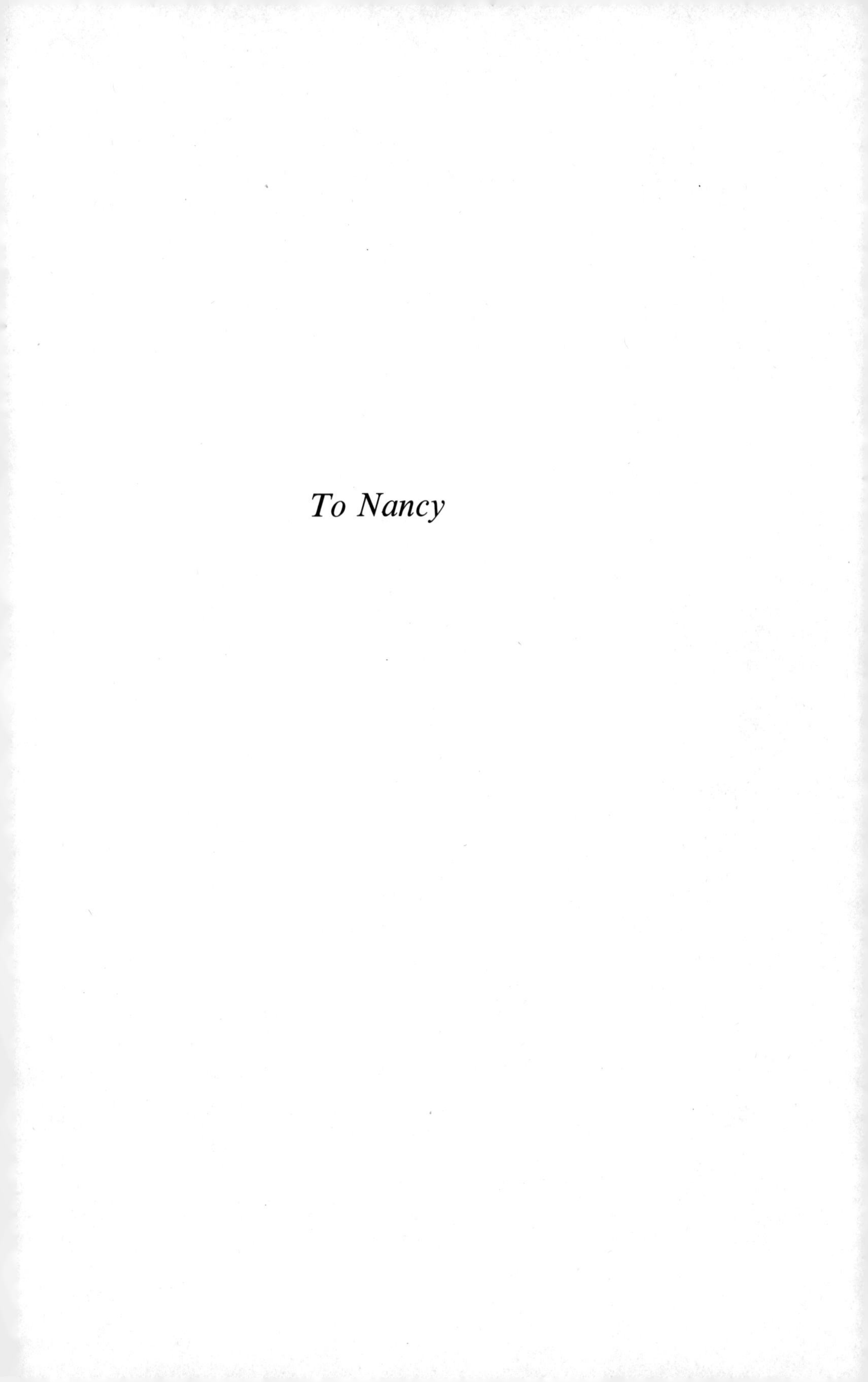

To Nancy

Preface

This book is an exposition of the technique of surgery on simply-connected smooth manifolds. Systematic study of differentiable manifolds using these ideas was begun by Milnor [45] and Wallace [68] and developed extensively in the last ten years. It is now possible to give a reasonably complete theory of simply-connected manifolds of dimension $\geqq 5$ using this approach and that is what I will try to begin here.

The emphasis has been placed on stating and proving the general results necessary to apply this method in various contexts. In Chapter II, these results are stated, and then applications are given to characterizing the homotopy type of differentiable manifolds and classifying manifolds within a given homotopy type. This theory was first extensively developed in Kervaire and Milnor [34] in the case of homotopy spheres, globalized by S. P. Novikov [49] and the author [6] for closed 1-connected manifolds, and extended to the bounded case by Wall [65] and Golo [23]. The thesis of Sullivan [62] reformed the theory in an elegant way in terms of classifying spaces.

Many applications have been omitted, such as applications to embedding theory [24], [38], [39], [25], [8], [9], [26], [27], study of manifolds with $\pi_1 = \mathbb{Z}$ [10], diffeomorphisms [11], and others. An exposition of applications to the theory of differentiable transformation groups is given in [12]. For a general discussion of surgery on non-simply-connected manifolds we refer to [66]. For extensions of the techniques to piecewise linear manifolds, we refer to [13] and [62]. In particular, the problem of computing with the classifying spaces for the *PL* theory has been now very well dealt with by Sullivan, and the recent work of Kirby and Siebenmann has shown how to extend all these results to topological manifolds. Discussion of these and many other beautiful developments are beyond the scope of this work, but I have tried here to introduce some of the basic ideas in the area of surgery, whose latest developments are so much involved with many of the most striking recent results in topology. A short exposition of some of the later developments can be found in my expository article "Manifolds and homotopy theory" in the Proceedings of the Amsterdam Conference on Manifolds, 1970, published by Springer.

The order of the chapters will not suit every taste. In particular, much of the contents of Chapter I will be quite familiar to many, and many readers will find more pleasure and motivation in beginning with Chapter II, and using Chapter I as a reference. The main ideas and results of surgery are in Chapter II while Chapter I develops some necessary tools in the theory of Poincaré complexes. Chapter III is an account of the simply-connected surgery obstruction, the index and Kervaire (Arf) invariant. Here, we have been forced to quote some rather difficult facts from the theory of integral quadratic forms, but we have developed everything needed in the theory of $\mathbb{Z}_2$-forms. The treatment of the Kervaire invariant is based on [7], and we include a treatment of product formuli in § 5 of Chapter III. Chapter IV proves the main theorem of surgery on 1-connected manifolds, following generally the point of view of [34]. In Chapter V we discuss "plumbing", which would have appeared in part II of [34].

In a later paper I hope to give a unified account of the applications of surgery to the study of submanifolds and "supermanifolds", based on the point of view of this book, (compare [8], [9], [10]).

This book was written partially at Princeton University and partially while the author was visiting at the Faculté des Sciences at Orsay of the University of Paris, and is based on courses given at Princeton 1966–1967, and at Orsay 1967–1968. I should like to thank also the Institute des Hautes Etudes Scientifiques, for their kind hospitality during that year, and the Mathematical Institute of the University of Warwick.

I am much indebted to many who made helpful comments and pointed out small mistakes, in particular, to David Singer, D.B.A. Epstein, Steven Weintraub, Michael Davis, and William Pardon. I was partially supported by the NSF while this work was under way.

Table of Contents

I. Poincaré Duality

In this chapter we will develop the properties of Poincaré duality spaces and pairs which play such an important role in the study of manifolds.

We begin in § 1 by studying the products, (slant, cup and cap) which relate homology and cohomology theories. In § 2 we study Poincaré duality in chain complexes and develop in this algebraic context the results needed for studying spaces or pairs, (such as compact manifolds or manifolds with boundary) for which Poincaré duality holds. We call them Poincaré complexes and pairs. In particular we study kernels and cokernels associated with maps of degree 1. In § 3 we study special forms of Poincaré duality, such as that for a bounded manifold with two pieces of boundary, and use these results to define the sums of Poincaré pairs and maps of degree 1. Then we use these results to prove Poincaré duality for smooth manifolds. In § 4 we discuss the Spivak normal fibre space of a Poincaré complex or pair, and prove Spivak's theorems on their existence and uniqueness.

Note that all chain complexes will be assumed free over $\mathbb{Z}$ in each dimension.

§ 1. Slant Operations, Cup and Cap Products

Let C be a chain complex, $C=\sum_{i\geqq 0} C_i, \partial: C_i \to C_{i-1}$. Let $C^* = \sum_{i\geqq 0} C^{-i}$, $C^{-i}=\operatorname{Hom}(C_i,\mathbb{Z})$ be the dual (cochain) complex, where $\delta: C^{-i}\to C^{-i-1}$ is defined by $\delta c=(-1)^i c\partial \in C^{-i-1}$ if $c\in C^{-i}$. $H^k(C)=H_{-k}(C^*)$.

If C, C' are complexes, define $C\otimes C'$ by $(C\otimes C')_n = \sum_{i+j=n} C_i\otimes C_j$ and $\partial(c\otimes c')=\partial c\otimes c' + (-1)^k c\otimes \partial c'$, if $c\in C_k$.

Define a map called the *slant operation*

$$/ : (C\otimes C')_n \otimes (C')^{-k} \to C_{n-k}$$

by the formula $a/b=\Sigma\, b(a_i')a_i$ where $a=\Sigma\, a_i\otimes a_i' \in C\otimes C'$, and $b(x)=0$ if $\dim x \neq k$.

Now we note that the slant operation is a chain map.

For we have $\partial a = \Sigma\, \partial a_i \otimes a'_i + (-1)^{d_i} a_i \otimes \partial a'_i$ if $\dim a_i = d_i$, so that

$$(\partial a)/b = \Sigma\, b(a'_i)\partial a_i + (-1)^{d_i} b(\partial a'_i) a_i .$$

Then $b(\partial a'_i) = 0$ unless $\dim \partial a'_i = k$ so $d_i = n-k-1$ and we get

$$\begin{aligned}(\partial a)/b &= \partial(a/b) + (-1)^{n-k-1} a/(b\partial)\\ &= \partial(a/b) + (-1)^{n-k-1} a/(-1)^k \delta b\\ &= \partial(a/b) + (-1)^{n-1} a/\delta b .\end{aligned}$$

So $\partial(a/b) = (\partial a)/b + (-1)^n a/\delta b$ and / is a chain map.

Since / is a chain map, it induces a map $H_*((C \otimes C') \otimes C'^*) \to H_*(C)$. Composing with the natural map

$$H_*(C \otimes C') \otimes H^*(C') \to H_*((C \otimes C') \otimes C'^*)$$

we get the slant operation

$$/ : H_n(C \otimes C') \otimes H^k(C') \to H_{n-k}(C) .$$

Now let C_i be an augmented chain complex with a diagonal map $\Delta : C \to C \otimes C$ such that $(\varepsilon \otimes 1)\Delta(c) = (1 \otimes \varepsilon)\Delta(c) = c$, where $\varepsilon : C \to \mathbb{Z}$ is the augmentation, and we identify $C = C \otimes \mathbb{Z} = \mathbb{Z} \otimes C$. Then Δ induces on C^* the structure of a ring with unit by $C^* \otimes C^* \to (C \otimes C)^* \xrightarrow{\Delta^*} C^*$, where $C^* \otimes C^* \to (C \otimes C)^*$ is the obvious inclusion. This is called the *cup product*, and on the cohomology level induces $H^*(C) \otimes H^*(C) \to H^*(C)$, also called the cup product, $x \otimes y \mapsto x \cup y$,

Using Δ, we may define the *cap product*

$$\cap : C_n \otimes C^{-k} \to C_{n-k}$$

by the formula $a \cap b = (\Delta a)/b$. Since $\cap$ is the composite of chain maps Δ and /, it follows that $\cap$ is a chain map and induces

$$\cap : H_n(C) \otimes H^k(C) \to H_{n-k}(C) .$$

More generally suppose A, B, and C are augmented chain complexes and let $\Delta : A \to B \otimes C$ be a chain map. Then similarly to the above we get a pairing in cohomology

$$\cup : H^p(B) \otimes H^q(C) \to H^{p+q}(A)$$

and a cap product

$$\cap : H_n(A) \otimes H^k(C) \to H_{n-k}(B) .$$

I.1.1 Proposition. *Let $x \in A_n$, $y \in C^{-k}$, $z \in B^{n-k}$. Then $z(x \cap y) = (z \cup y)(x)$.*

Proof. $x \cap y = (\Delta x)/y = \Sigma\, y(x'_i) x_i$ where $\Delta x = \Sigma\, x_i \otimes x'_i$, $x_i \in B$, $x'_i \in C$. Then $z(x \cap y) = \Sigma\, y(x'_i) z(x_i) = (z \otimes y)(\Sigma\, x_i \otimes x'_i) = (z \otimes y)(\Delta x) = \Delta^*(z \otimes y)(x)$ $= (z \cup y)(x)$. □

I.1.2 Corollary. *Suppose we have a commutative diagram:*

$$\begin{array}{ccc} A & \xrightarrow{\Delta_1} & B' \otimes C \\ {\scriptstyle \Delta_2}\downarrow & & \downarrow{\scriptstyle \Delta_4 \otimes 1} \\ B \otimes C' & \xrightarrow{1 \otimes \Delta_3} & B \otimes D \otimes C\,. \end{array}$$

If $x \in A_n$, $y \in C^{-k}$, $z \in D^{-q}$, *then*

$$(x \cap y) \cap z = x \cap (z \cup y) \in B_{n-k-q},$$

where $x \cap y \in B'_{n-k}$, $z \cup y \in C'^{-k-q}$, *cup and cap products being defined by the appropriate diagonal* Δ_i, $i = 1, 2, 3$ *or* 4, *in each case.*

Proof. Let $w \in B^{-(n-k-q)}$. By (I.1.1),

$$w(x \cap (z \cup y)) = (w \cup (z \cup y))(x),$$

and

$$\begin{aligned} w \cup (z \cup y) &= \Delta_2^*(w \otimes \Delta_3^*(z \otimes y)) = \Delta_2^*(1 \otimes \Delta_3^*)(w \otimes z \otimes y) \\ &= \Delta_1^*(\Delta_4^* \otimes 1)(w \otimes z \otimes y) = (w \cup z) \cup y, \end{aligned}$$

and

$$((w \cup z) \cup y)(x) = (w \cup z)(x \cap y) = w((x \cap y) \cap z), \text{ by (I.1.1)}.$$

Hence

$$w(x \cap (z \cup y)) = w((x \cap y) \cap z)$$

for any

$$w \in B^{-(n-k-q)}, \quad \text{so} \quad x \cap (z \cup y) = (x \cap y) \cap z. \quad \square$$

I.1.3 Corollary. *Let* $f: A \to A'$, $g: B \to B'$, $h: C \to C'$ *be chain maps such that*

$$\begin{array}{ccc} A & \xrightarrow{\Delta} & B \otimes C \\ {\scriptstyle f}\downarrow & & \downarrow{\scriptstyle g \otimes h} \\ A' & \xrightarrow{\Delta'} & B' \otimes C' \end{array}$$

commutes.

Then $g_*(x \cap h^* y') = (f_*(x)) \cap y'$, *for* $x \in H_*(A)$, $y' \in H^*(C')$.

Proof. By (I.1.1), if $z' \in B'^*$, then

$$\begin{aligned} z'(g_*(x \cap h^* y')) &= (g^*(z'))(x \cap h^* y') = (g^* z' \cup h^* y')(x) \\ &= (f^*(z' \cup y'))(x) = (z' \cup y')(f_*(x)) = z'(f_*(x) \cap y'). \end{aligned}$$

Hence $g_*(x \cap h^* y') = (f_*(x)) \cap y'$. $\square$

Let $0 \to A \xrightarrow{i} B \xrightarrow{j} C \to 0$ be an exact sequence of chain maps, where $i: A \to B$ is a map of chain complexes with diagonal maps, so

$$\begin{array}{ccc} A & \xrightarrow{i} & B \\ \downarrow \Delta & & \downarrow \Delta \\ A \otimes A & \xrightarrow{i \otimes i} & B \otimes B \end{array}$$

commutes. Then we may define $\Delta_1 : C \to B \otimes C$, by $\Delta_1(c) = (1 \otimes j)(\Delta b)$, where $b \in B$ and $jb = c$. If $jb' = jb = c$ then $b' = b + a$, $a \in A$ and $\Delta b' = \Delta b + \Delta a$, where $\Delta a \in A \otimes A$. Hence $(1 \otimes j)(\Delta a) = 0$ and Δ_1 is well defined. Similarly define $\Delta_2(c) = (j \otimes 1)\Delta(b)$, and this is also well defined, $\Delta_2 : C \to C \otimes B$.

Then Δ_1 and Δ_2 define cup and cap products, in particular,

$$\cap : H_n(C) \otimes H^k(C) \to H_{n-k}(B)$$
$$\cap : H_n(C) \otimes H^q(B) \to H_{n-q}(C) .$$

Let $\partial : H_q(C) \to H_{q-1}(A)$ and $\delta : H^{k-1}(A) \to H^k(C)$ be the boundary and coboundary operators associated with the exact sequences.

I.1.4 Proposition. *Let $x \in H_n(C)$, $y \in H^k(C)$, $z \in H^q(B)$, $u \in H^{k-1}(A)$. Then*

(i) $j_*(x \cap y) = x \cap j^* y$,

(ii) $\partial(x \cap z) = (\partial x) \cap (i^* z)$,

(iii) $(-1)^{n-1} x \cap (\delta u) = i_*(\partial x \cap u)$.

Proof. Let jc be a chain representing x, $c \in B$, $b \in C^{-k}$ representing y. Then $x \cap y$ is represented by $((1 \otimes j)\Delta c)/b = \Sigma\, c_i \otimes jc_i'/b = \Sigma\, b(jc_i')c_i$. Then $j_*(x \cap y)$ is represented by $\Sigma\, b(jc_i')jc_i = \Sigma\, ((bj)(c_i'))jc_i = (j \otimes 1)\Delta c/bj$ which represents $x \cap j^* y$. This proves (i).

To prove (ii) we first recall the definition of $\partial : H_n(C) \to H_{n-1}(A)$. If $x \in H_n(C)$ is represented by a chain $jc \in C$, $\partial c = ir$, and ∂x is represented by the chain r. Let $\Delta r = \Sigma\, r_i \otimes r_i'$, $\Delta c = \Sigma\, c_i \otimes c_i'$. If $b \in B^{-q}$ is a cocycle representing $z \in H^q(B)$, then $\partial x \cap i^* z$ is represented by

$$\Delta r/i^* b = \Delta r/bi = \Sigma\, (bi(r_i'))r_i .$$

Then

$$i(\Delta r/bi) = \Sigma\, (bi(r_i'))ir_i = i(\Delta r)/b = \Delta \partial c/b = \partial \Delta c/b = \partial(\Delta c/b), \text{ since } \delta b = 0.$$

But $x \cap z$ is represented by $(j \otimes 1)\Delta c/b = j(\Delta c/b)$ so that $\partial(x \cap z)$ is represented by $a \in A$ such that $ia = \partial(\Delta c/b)$. Hence $i(a) = i(\Delta r/bi)$, and since i is mono, $a = \Delta r/bi$ and $\partial(x \cap z) = \partial x \cap i^* z$.

In (iii), let $r \in A^*$ be such that r represents $u \in H^{k-1}(A)$, let $s \in B^*$ so that $r = i^* s = si$, and let $t \in C^*$ so that $\delta s = (-1)^{k-1} s\partial = j^* t = tj$. Then t represents $\delta u \in H^k(C)$. If $c \in B_n$ such that jc represents $x \in H_n(C)$, and

$a \in A_{n-1}$ such that $ia = \partial c$ then a represents $\partial x \in H_{n-1}(A)$. Suppose $\Delta a = \Sigma a_i \otimes a_i'$ so that $i \otimes i \Delta a = \Delta i a = \Delta \partial c$, and $\Delta c = \Sigma c_i \otimes c_i'$. Then $i_*(\partial x \cap u)$ is represented by

$$i(\Delta a/r) = i(\Sigma r(a_i') a_i) = \Sigma r(a_i') i(a_i) = \Sigma (si(a_i')) i(a_i) = (\Delta \partial c)/s = (\partial \Delta c)/s.$$

Since / is a chain map, we have

$$\partial(\Delta c/s) = (\partial \Delta c)/s + (-1)^n \Delta c/\delta s.$$

On the other hand

$$\Delta c/\delta s = \Delta c/j^* t = (j \Delta c)/t = (\Delta j c)/t$$

so that $\Delta c/\delta s$ represents $x \cap (\delta u)$. Since $\partial(\Delta c/s)$ is a boundary in B, the homology classes of $(\partial \Delta c)/s$ and $(-1)^{n-1} \Delta c/\delta s$ are the same and (iii) follows. □

I.1.5 Theorem. *Let $0 \to A \xrightarrow{i} B \xrightarrow{j} C \to 0$ be exact, $x \in H_m(C)$, i a map of chain complexes with diagonal. Then*

$$\begin{array}{ccccccccc}
\cdots \longrightarrow H^q(C) & \xrightarrow{j^*} & H^q(B) & \xrightarrow{i^*} & H^q(A) & \xrightarrow{\delta} & H^{q+1}(C) \longrightarrow \cdots \\
\downarrow {\scriptstyle \cap x} & & \downarrow {\scriptstyle \cap x} & & \downarrow {\scriptstyle \cap \partial x} & & \downarrow {\scriptstyle \cap x} \\
\cdots \longrightarrow H_{m-q}(B) & \xrightarrow{j_*} & H_{m-q}(C) & \xrightarrow{\partial} & H_{m-q-1}(A) & \xrightarrow{i_*} & H_{m-q-1}(B) \longrightarrow \cdots
\end{array}$$

is commutative, up to sign.

Proof. Let $y \in H^q(C)$. Then $x \cap j^* y = j_*(x \cap y)$ by (I.1.4) (i). If $y \in H^q(B)$, then $\partial(x \cap y) = (\partial x) \cap i^* y$ by (I.1.4) (ii). If $u \in H^q(A)$, then

$$(-1)^{n-1} x \cap \delta u = i_*(\partial x \cap u),$$

by (I.1.4) (iii). Hence the first two squares commute and the third commutes up to sign. □

I.1.6 Proposition. *Suppose $0 \to A \xrightarrow{i} B \xrightarrow{j} C \to 0$ is an exact sequence of chain complexes, where $A = A' + A''$, the sum (not necessarily disjoint) of augmented chain complexes with diagonal map, and $i: A \to B$ is a map of augmented chain complexes with diagonal maps. Let $C' = B/A'$, $C'' = B/A''$. Then there is a natural diagonal $\Delta_0: C \to C' \otimes C''$ with the following properties: Let $j': B \to C'$, $j'': B \to C''$, $\eta': C' \to C$, $\eta'': C'' \to C$ be the natural maps. Let $x \in H_n(C)$, $y \in H^k(C)$, $z \in H_n(B)$, $u \in H^q(C'')$. Then*

(i) $x \cap j''^* u = \eta'_*(x \cap u) \in H_{n-q}(C)$,
(ii) $j'_*(z \cap j''^* u) = j_* z \cap u \in H_{n-q}(C')$,
(iii) $j'_*(x \cap y) = x \cap \eta''^* y \in H_{n-k}(C')$.

The proof consists of routine chain arguments as in (I.1.4) and we omit them.

Now we bring in spaces.

I.1.7 Theorem. *If $C = C_*(X)$, the singular chains of a topological space X, and $\Delta : C \to C \otimes C$ is the diagonal induced by $d : X \to X \times X$, $d(x) = (x, x)$ and the Eilenberg-Silber map (see [21])*

$$C_*(X \times X) \to C_*(X) \otimes C_*(X),$$

then all the results of this section hold for the various products induced by Δ.

The proof is trivial.

Analogous to the results on cap products, we may deduce similar properties of the various cup products in these situations. These may be deduced directly, or using the results on cap products and (I.1.2).

I.1.8 Lemma. *Let $C \to C' \otimes C''$ be a diagonal C', C'' complexes with locally finitely generated homology and, let $c \in C_n$, $\{c\} = v \in H_n(C)$. Then $v \cap : H^q(C'') \to H_{n-q}(C')$ is an isomorphism for all q, if and only if the pairing induced by cup product $\psi : H^{n-q}(C' \otimes \mathbb{Z}_p) \otimes H^q(C'' \otimes \mathbb{Z}_p) \to \mathbb{Z}_p$, $\psi(x \otimes y) = (x \cup y)(v)$ is non-singular for each prime p, for all q.*

Proof. $c \cap : C''^* \to C'$ is a chain map inducing $v \cap$ on the homology level. Then $v \cap$ is an isomorphism if and only if the homology of the mapping cone M of the chain map $c \cap$, is zero, (see [22; V § 13]). Since C''^* and C' have locally finitely generated homology, it follows that M has locally finitely generated homology. The Universal Coefficient Theorem (see [22; p. 161]) then shows that $H_*(M) = 0$ if and only if $H_*(M \otimes \mathbb{Z}_p) = 0$ for all primes p. But $M \otimes \mathbb{Z}_p$ is the mapping cone of $c \cap : C''^* \otimes \mathbb{Z}_p \to C' \otimes \mathbb{Z}_p$, so $H_*(M \otimes \mathbb{Z}_p) = 0$ if and only if

$$v \cap : H^q(C'' \otimes \mathbb{Z}_p) \to H_{n-q}(C' \otimes \mathbb{Z}_p)$$

is an isomorphism for all q. But $H^{n-q}(C' \otimes \mathbb{Z}_p) = \operatorname{Hom}(H_{n-q}(C' \otimes \mathbb{Z}_p), \mathbb{Z}_p)$. Hence $v \cap : H^q(C'' \otimes \mathbb{Z}_p) \to H_{n-q}(C' \otimes \mathbb{Z}_p)$ is an isomorphism if and only if the pairing $H^{n-q}(C' \otimes \mathbb{Z}_p) \otimes H^q(C'' \otimes \mathbb{Z}_p) \to \mathbb{Z}_p$ given by $x \otimes y \to x(v \cap y)$ is non-singular. But $x(v \cap y) = (x \cup y)(v)$ by (I.1.1). □

§ 2. Poincaré Duality

Since most of the results we give here are of a purely algebraic nature, we will state them in the algebraic context of chain complexes. All statements translate immediately into topological ones, taking the chain complex of a space.

All chain complexes C considered in § 2 will be assumed to have locally finitely generated homology, i.e. $H_i(C)$ is a finitely generated

$\mathbb{Z}$-module for each i, and also we will assume $C_i = 0$ for $i < 0$. A *geometric chain complex* will mean an augmented chain complex C with a diagonal map $\Delta: C \to C \otimes C$ and a chain homotopy H between Δ and $T\Delta$, so that $\partial H + H\partial = \Delta - T\Delta$, where $T(a \otimes b) = (-1)^{\varepsilon} b \otimes a$, $\varepsilon = (\dim a)(\dim b)$. A *geometric chain map* will be a chain map $f: C \to C'$ where C, C' are geometric complexes, such that $\Delta' f = (f \otimes f)\Delta$ and $H'f = (f \otimes f)H$. A *geometric chain pair* (B, A) will be a geometric chain complex B with a subcomplex A, which is a direct summand as a graded module and a geometric chain complex, such that the inclusion $A \subset B$ is a geometric chain map. The chain complex of a space is the prime example of a geometric chain complex. We also denote by (B, A) the free chain complex B/A.

A geometric chain complex C will be called a *Poincaré chain complex* of dimension n, if there exists $\mu \in H_n(C)$ of infinite order such that $\mu \cap: H^k(C) \to H_{n-k}(C)$ is an isomorphism for each k.

A geometric chain pair (B, A) will be called a *Poincaré chain pair* of dimension m if there is an element $v \in H_m(B, A) = H_m(B/A)$ of infinite order such that $v \cap: H^q(B) \to H_{m-q}(B, A)$ is an isomorphism for all q. The element v (or μ) is called the *orientation class* of (B, A) (or of C), and the choice of v (or μ) is called an *orientation* of (B, A) (or of C). If (X, Y) is a pair of CW complexes which satisfies Poincaré duality (i.e. whose chain complex is a Poincaré chain pair) then we call (X, Y) a *Poincaré pair*, while if $Y = \emptyset$ so X satisfies Poincaré duality, we call X a *Poincaré complex*.

I.2.1 Proposition. *Let (B, A) be a geometric chain pair such that $H_n(B, A) = \mathbb{Z}$ with generator v. Then the following three conditions are equivalent:*

(a) $v \cap: H^q(B) \to H_{m-q}(B, A)$ *is an isomorphism for each* q.

(b) $v \cap: H^q(B, A) \to H_{m-q}(B)$ *is an isomorphism for each* q.

(c) *The pairing* $\psi: H^q(B; \mathbb{Z}_p) \otimes H^{m-q}(B, A; \mathbb{Z}_p) \to \mathbb{Z}_p$ *given by* $\psi(x, y) = (x \cup y)(v)$, *is non-singular for every prime* p, *each* q.

Proof. By (I.1.8) (b) and (c) are equivalent. By (I.1.8) (a) is equivalent to the statement that the pairing.

$$\psi': H^q(B, A; \mathbb{Z}_p) \otimes H^{m-q}(B \otimes \mathbb{Z}_p) \to \mathbb{Z}_p, \quad \psi'(y \otimes x) = (y \cup x)(v),$$

is non-singular for all primes p, all q.

Since (B, A) is a geometric chain pair there is a chain homotopy H between Δ and $T\Delta$ and $H(A) \subset A \otimes A$. Hence H induces a chain homotopy between $\Delta': C \to C \otimes B$ and $T\Delta'': C \to C \otimes B$, where $C = B/A$. It follows that $\psi'(y \otimes x) = (-1)^{q(m-q)} \psi(x \otimes y)$. Hence ψ is non-singular if and only if ψ' is non-singular, and the Proposition follows. □

I.2.2 Theorem. *If (B, A) is a Poincaré chain pair of dimension m, then the diagram commutes up to sign:*

$$\begin{array}{ccccccc}
\to H^q(B,A) & \xrightarrow{j^*} & H^q(B) & \xrightarrow{i^*} & H^q(A) & \xrightarrow{\delta} & H^{q+1}(B,A)\to \\
\downarrow{\scriptstyle \nu\cap} & & \downarrow{\scriptstyle \nu\cap} & & \downarrow{\scriptstyle (\partial\nu)\cap} & & \downarrow{\scriptstyle \nu\cap} \\
\to H_{m-q}(B) & \xrightarrow{j_*} & H_{m-q}(B,A) & \xrightarrow{\partial} & H_{m-q-1}(A) & \xrightarrow{i_*} & H_{m-q-1}(B)\to
\end{array}$$

and all the vertical arrows are isomorphisms.

Proof. By (I.1.5), the diagram commutes up to sign. Since (B, A) is a Poincaré chain pair, $\nu\cap : H^q(B)\to H_{m-q}(B, A)$ is an isomorphism for all q. Hence by (I.2.1), $\nu\cap : H^q(B, A)\to H_{m-q}(B)$ is an isomorphism for all q. Then by the Five Lemma $(\partial\nu)\cap : H^q(A)\to H_{m-1-q}(A)$ is an isomorphism for all q. □

I.2.3 Corollary. *If (B, A) is a Poincaré chain pair of dimension m, then A is a Poincaré chain complex of dimension $m-1$.*

Proof. By (I.2.2), $(\partial\nu)\cap : H^q(A)\to H_{m-q-1}(A)$ is an isomorphism for all q, so it remains to check only that $\partial\nu$ is of infinite order in $H_{m-1}(A)$. However, if $N(\partial\nu)=0$, some N, then $N(\partial\nu\cap x)=0, x\in H^*(A)$, so that $NH_*(A)=0$. But A is an augmented complex so that $\mathbb{Z}\in H_0(A)$, and hence $N(\partial\nu)\neq 0$ for all N. □

We will make a convention that a Poincaré chain pair (B, A) where $A=0$ will mean a Poincaré chain complex B.

If ν is the orientation class of the Poincaré chain pair (B, A), then $\partial\nu$ will be the orientation class of A, by convention ("compatibly oriented").

If (B, A) and (B', A') are oriented Poincaré chain pairs of dimension m, a chain map $f : (B, A)\to(B', A')$ will be said to have degree 1 if $f_*(\nu)=\nu'$, where ν, ν' are the orientation classes of (B, A) and (B', A') respectively, where $f_* : H_*(B, A)\to H_*(B', A')$. We denote the induced map on $H_*(B)$ by $\bar{f}_* : H_*(B)\to H_*(B')$, and similar notation in cohomology.

I.2.4 Lemma. *If $f : (B, A)\to(B', A')$ is a map of degree 1, then $f'=f\,|\,A : A\to A'$ is a map of degree 1.*

Proof. $f_*(\nu)=\nu'$, so $f_*(\partial\nu)=\partial f_*(\nu)=\partial\nu'$. But $\partial\nu$ and $\partial\nu'$ are the orientation classes of A and A'. □

I.2.5 Theorem. *Maps of degree 1 split, i.e. with notation as above, there exist*

$$\alpha_* : H_*(B', A')\to H_*(B, A),\ \beta_* : H_*(B')\to H_*(B),$$

$$\alpha^* : H^*(B, A)\to H^*(B', A'),\ \beta^* : H^*(B)\to H^*(B')$$

such that $f_*\alpha_*=1, \bar{f}_*\beta_*=1, \alpha^*f^*=1, \beta^*\bar{f}_*=1.$

Proof. Let $P: H_{m-q}(B', A') \to H^q(B')$ and $\bar{P}: H_{n-q}(B') \to H^q(B', A')$ be the inverses of the Poincaré duality maps, so that $v' \cap P(x) = x$, $v' \cap \bar{P}(y) = y$, $x \in H_{m-q}(B', A')$, $y \in H_{m-q}(B')$. We define:

$$\alpha_*(x) = v \cap \bar{f}^*(P(x)),\ x \in H_{m-q}(B', A')$$
$$\beta_*(y) = v \cap f^*(\bar{P}(y)),\ y \in H_{m-q}(B')$$
$$\alpha^*(u) = \bar{P}(\bar{f}_*(v \cap u)),\ u \in H^q(B, A)$$
$$\beta^*(v) = P(f_*(v \cap v)),\ v \in H^q(B).$$

Using (I.1.3), we have

$$f_* \alpha_*(x) = f_*(v \cap \bar{f}^*(P(x))) = f_*(v) \cap P(x) = v' \cap P(x) = x.$$

Similarly

$$\bar{f}_* \beta_*(y) = \bar{f}_*(v \cap f^*(\bar{P}(y))) = v' \cap \bar{P}(y) = y.$$

Also

$$v' \cap \alpha^* f^*(z) = v' \cap \bar{P}(\bar{f}_*(v \cap f^*(z))) = \bar{f}_*(v \cap f^*(z)) = v' \cap z,$$

for $z \in H^q(B', A')$. Since $v' \cap$ is an isomorphism, $\alpha^* f^*(z) = z$. Similarly, one shows $\beta^* \bar{f}^*(w) = w$ for $w \in H^q(B')$. □

It follows from (I.2.5), that there are direct sum splittings

$$H_*(B, A) = \ker f_* + \operatorname{image} \alpha_*,\ H_*(B) = \ker \bar{f}_* + \operatorname{im} \beta_*,$$
$$H^*(B, A) = \operatorname{im} f^* + \ker \alpha^*,\ H^*(B) = \operatorname{im} \bar{f}^* + \ker \beta^*.$$

Let us establish the following notation that will be used throughout this book. Let

$$K_q(B, A) = (\ker f_*)_q \subset H_q(B, A),$$
$$K_q(B) = (\ker \bar{f}_*)_q \subset H_q(B),$$
$$K^q(B, A) = (\ker \alpha^*)^q \subset H^q(B, A),$$
$$K^q(B) = (\ker \beta^*)^q \subset H^q(B),$$
$$K_q(B, A; G) = (\ker f_*)_q \subset H_q(B, A; G), \quad \text{etc.}$$

Then we may derive the following properties of K_q and K^q:

I.2.6. *$v \cap$ preserve the direct sum splitting, so*

$$v \cap K^q(B, A) \subset K_{m-q}(B), \quad v \cap K^q(B) \subset K_{m-q}(B, A)$$

and $v \cap : K^q(B, A) \to K_{m-q}(B)$ and $v \cap : K^q(B) \to K_{m-q}(B, A)$ are isomorphisms, ("K^* and K_* satisfy Poincaré duality").

I.2.7. *In the exact homology and cohomology sequences of* (B, A), *all the maps preserve the direct sum splitting, so induce a diagram, com-*

mutative up to sign, with exact rows:

$$\begin{array}{ccccccccc}
\cdots \xrightarrow{i^*} & K^{q-1}(A) & \xrightarrow{\delta} & K^q(B,A) & \xrightarrow{j^*} & K^q(B) & \xrightarrow{i^*} & K^q(A) & \xrightarrow{\delta} \cdots \\
& \downarrow{\scriptstyle \partial v\cap} & & \downarrow{\scriptstyle v\cap} & & \downarrow{\scriptstyle v\cap} & & \downarrow{\scriptstyle \partial v\cap} & \\
\cdots \xrightarrow{\partial} & K_{m-q}(A) & \xrightarrow{i_*} & K_{m-q}(B) & \xrightarrow{j_*} & K_{m-q}(B,A) & \xrightarrow{\partial} & K_{m-q-1}(A) & \xrightarrow{i_*} \cdots
\end{array}$$

(*In particular we have relations*

$$i_*\gamma_* = \alpha_* i'_*,\ \gamma_* : H_q(A') \to H_q(A),\ i' : A' \to B',\ etc.)$$

I.2.8. *The Universal Coefficient Formulas hold for* K^* *and* K_*, i.e.

$$K_q(B, A; G) = K_q(B, A) \otimes G + \operatorname{Tor}(K_{q-1}(B, A), G)$$

$$K_q(B; G) = K_q(B) \otimes G + \operatorname{Tor}(K_q(B), G)$$

$$K^q(B, A; G) = \operatorname{Hom}(K_q(B, A), G) + \operatorname{Ext}(K_{q-1}(B, A), G)$$

$$K^q(B, G) = \operatorname{Hom}(K_q(B), G) + \operatorname{Ext}(K_{q-1}(B), G)\,.$$

It is useful to have following interpretation of (I.2.6):

I.2.9. *Under the pairing* $H^q(B; F) \otimes H^{m-q}(B, A; F) \to F$, *given by* $(x, y) = (x \cup y)(y)$, *(F a ring),* $K^{m-q}(B, A; F)$ *is orthogonal to* $\bar{f}^*(H^q(B'; F))$, $K^q(B; F)$ *is orthogonal to* $f^*(H^{m-q}(B', A'; F))$, *and on*

$$K^q(B; F) \otimes K^{m-q}(B, A; F)$$

the pairing is non-singular if F is a field. If $F = \mathbb{Z}$ *it is non-singular on* $K^q(B)$/torsion $\otimes$ $K^{m-q}(B, A)$/torsion.

Proof of (I.2.6). Let $u \in K^q(B, A)$. Then $\alpha^*(u) = 0$, and $\alpha^*(u) = \bar{P}(\bar{f}_*(v \cap u))$. Since $\bar{P}$ is an isomorphism, $\bar{f}_*(v \cap u) = 0$, and $v \cap u \in K_{m-q}(B)$. Also $v \cap f^*(z) = v \cap f(\bar{P}(v' \cap z)) = \beta_*(v' \cap z)$, so $v \cap \operatorname{im} f^* \subset \operatorname{im} \beta_*$, and $v\cap$ preserves one of the direct sum splittings. The other follows in a similar way. Then, since $v\cap$ is an isomorphism, it follows that each summand is mapped isomorphically.

Proof of (I.2.7). Using Poincaré duality (I.2.6), it suffices to show that the homology maps preserve the summands in the splittings. Since K_q are defined as kernels of the homology maps, they are clearly preserved, so i_*, j_* and ∂ send K_* into K_*.

Denote by $\gamma_* : H_q(A') \to H_q(A)$ the splitting of $f'_* = (f \mid A)_*$, and $P' : H_{m-q-1}(A') \to H^q(A')$, the inverse of Poincaré duality.

Let $z \in H_q(A')$, so $\gamma_*(z) = \partial v \cap f'^*(P'z)$. Then by (I.1.4) (iii), $i_*(\gamma_*(z)) = (-1)^{m-1} v \cap \delta f'^*(P'z)$. Then $\delta f'^* = f^*\delta$, and $(-1)^{m-1} v \cap (\delta P'z) = i'_*(\partial v' \cap P'z) = i'_*(z)$. Hence $\delta P'(z) = \bar{P}(i'_*(z))$ and

$$i_*(\gamma_*(z)) = (-1)^{m-1} v \cap f^*(\delta P'(z)) = (-1)^{m-1} v \cap f^* \bar{P}(i'_* z) = (-1)^{m-1} \alpha_*(i'_* z)\,,$$

so $i_*\gamma_* = (-1)^{m-1}\alpha_* i'_*$, and i_* preserves the direct sum splitting, i.e., $i_*(\operatorname{im}\gamma_*) \subset \operatorname{im}\alpha_*$.

Let $y \in H_q(B')$, so $\beta_*(y) = v \cap f^*(\bar{P}(y))$. Then by (I.1.4) (i),

$$j_* \beta_*(y) = j_*(v \cap f^*(\bar{P}(y))) = v \cap j^* f^*(\bar{P}(y)) = v \cap \bar{f}^{\,*} j'^* \bar{P}(y).$$

Now

$$v' \cap j'^* \bar{P} y = j'_*(v' \cap \bar{P} y) = j'_*(y), \quad \text{so} \quad j'^* \bar{P}(y) = P j'_* y.$$

Hence $j_* \beta_*(y) = v \cap \bar{f}^* P j'_* y = \alpha_* j'_* y$, so $j_* \beta_* = \alpha_* j'_*$, and j_* preserves the direct sum splitting, i.e. $j_*(\operatorname{im} \beta_*) \subset \operatorname{im} \alpha_*$.

Let $x \in H_q(B', A')$, so $\alpha_*(x) = v \cap \bar{f}^* P(x)$. Then by (I.1.4) (ii) $\partial \alpha_* x = \partial(v \cap \bar{f}^* P(x)) = \partial v \cap i^* \bar{f}^* P(x) = \partial v \cap f'^* i'^* P x$.

Now $i'^* P(x) = P' \partial x$, since by (I.1.4) (ii) $(\partial v) \cap i'^* P(x) = \partial(v' \cap P x) = \partial x$. Hence $\partial \alpha_* x = \partial v \cap f'^* P' \partial x = \gamma_* \partial x$. Therefore $\partial \alpha_* = \gamma_* \partial$, and $\partial(\operatorname{im} \alpha_*) \subset \operatorname{im} \gamma_*$, and ∂ preserves the direct sum splitting. □

Proof of (I.2.8). We have the exact sequence of the map f

$$\cdots \longrightarrow H_{q+1}(f) \xrightarrow{\ \partial\ } H_q(B, A) \xrightarrow{\ f_*\ } H_q(B', A') \longrightarrow \cdots$$

where $H_*(f)$ is the homology of the mapping cylinder of f, which is a free chain complex since (B, A) and (B', A') are free. Hence the universal coefficient formula holds for $H_*(f)$. But ∂ maps $H_{q+1}(f)$ isomorphically onto $K_q(B, A) = \ker f_*$, since f_* is split, and hence the Universal Coefficient formula holds for $K_q(B, A)$. Similar proofs hold in the other cases. □

Proof of (I.2.9). By (I.2.6),

$$(v \cap K^q(B, A; F)) \subset K_{m-q}(B; F) = (\ker \bar{f}_*)_{m-q},$$
$$\bar{f}_* : H_{m-q}(B; F) \longrightarrow H_{m-q}(B', F).$$

Then using (I.1.2), if $x \in K^q(B, A; F)$, $y' \in H^{m-q}(B'; F)$, we have

$$(\bar{f}^* y' \cup x)(v) = (\bar{f}^* y')(v \cap x) = y'(\bar{f}_*(v \cap x)) = 0$$

since $v \cap x \in K_{m-q}(B, F) = (\ker \bar{f}_*)_{m-q}$. Hence $K^q(B, A; F)$ is orthogonal to $\bar{f}^* H^*(B'; F)$. Similarly one may show $K^{m-q}(B; F)$ is orthogonal to $f^* H^{m-q}(B, A; F)$. But since the pairing is non-singular if F is a field on all of $H^q(B, A; F) \otimes H^{m-q}(B; F)$, it follows that the restriction to $K^q(B, A; F) \otimes K^{m-q}(B; F)$ is non-singular.

Now if $F = \mathbb{Z}$, and if $qx = 0$, $q \in \mathbb{Z}$, then $q(x, y) = (qx, y) = 0$ any y. Since the values of $(\,,\,)$ are in $\mathbb{Z}$, this implies that $(x, y) = 0$. Hence Torsion $K^q(B, A)$ annihilates $K^{m-q}(B)$ and Torsion $K^{m-q}(B)$ annihilates $K^q(B, A)$. Hence we have an induced pairing on

$$(K^q(B, A)/\text{torsion}) \otimes (K^{m-q}(B)/\text{torsion}) \longrightarrow \mathbb{Z}.$$

Tensoring over $\mathbb{Q}$, it follows from the result with $F = \mathbb{Q}$ that

$$K^q(B, A)/\text{torsion} \longrightarrow \operatorname{Hom}(K^{m-q}(B)/\text{torsion}, \mathbb{Z})$$

is a monomorphism. If it is not an isomorphism, then tensoring over $\mathbb{Z}_p$ for some prime p, it has a kernel. But the result with $F = \mathbb{Z}_p$ implies this is not the case, so the result follows with $F = \mathbb{Z}$. □

§ 3. Poincaré Pairs and Triads; Sums of Poincaré Pairs and Maps

In this section we will consider geometric chain pairs, all maps will be assumed to commute with the diagonal map and all inclusions of subcomplexes will be assumed to split, and commute with the chain homotopy of the diagonals. That is to say, in the notation of § 2, we assume all pairs involved are geometric chain pairs.

I.3.1 Proposition. *Let (B, A) be a geometric chain pair,*

$$A = A_1 + A_2, \ A_0 = A_1 \cap A_2 ,$$

where (A, A_i), and (A_i, A_0), $i = 1, 2$ are geometric chain pairs. Then there is a diagonal defined $\Delta_3 : A/A_2 \to A/A_2 \otimes A_1$ such that the diagram

$$\begin{array}{ccc} A_1/A_0 & \xrightarrow{\Delta_2} & A_1/A_0 \otimes A_1 \\ \downarrow & & \downarrow \\ A/A_2 & \xrightarrow{\Delta_3} & A/A_2 \otimes A_1 \end{array}$$

commutes, where $A_1/A_0 \to A/A_2$ is the isomorphism induced by the inclusion $(A_1, A_0) \to (A, A_2)$. Further the cap product defined has the following properties:

(i)
$$\begin{array}{ccc} H^q(B, A_2) & \xrightarrow{i^*} & H^q(A, A_2) \\ \downarrow{\scriptstyle \nu\cap} & & \downarrow{\scriptstyle \partial\nu\cap} \\ H_{m-q}(B, A_1) & \xrightarrow{\partial} & H_{m-q-1}(A_1) \end{array}$$
commutes,

(ii)
$$\begin{array}{ccc} H^{q-1}(A, A_2) & \xrightarrow{\delta} & H^q(B, A) \\ \downarrow{\scriptstyle \partial\nu\cap} & & \downarrow{\scriptstyle \nu\cap} \\ H_{m-q}(A_1) & \xrightarrow{i_*} & H_{m-q}(B) \end{array}$$
commutes up to $(-1)^{m-1}$,

(iii)
$$\begin{array}{ccc} H^q(B) & \xrightarrow{i^*} & H^q(A_1) \\ \downarrow{\scriptstyle \nu\cap} & & \downarrow{\scriptstyle (\partial\nu)\cap} \\ H_{m-q}(B, A) & \xrightarrow{\partial} & H_{m-q-1}(A, A_2) \end{array}$$
commutes,

where $\nu \in H_m(B, A)$ and $\partial : H_j(B, A) \to H_{j-1}(A, A_2)$ so $\partial\nu \in H_{m-1}(A, A_2)$.

The proof of (I.3.1) is routine and we omit it.

Suppose $B = B_1 + B_2$, $B_0 = B_1 \cap B_2$, $A_i = B_i \cap A$, all pairs are geometric chain pairs, so that we have chain isomorphisms:

$$B_1/(B_0 + A_1) \to B/(B_2 + A),$$
$$B_2/(B_0 + A_2) \to B/(B_1 + A),$$

(corresponding to excisions in the geometrical picture). Also we have Mayer-Vietoris sequences (see [22]).

$$\cdots \to H_{q+1}(B) \to H_q(B_0) \to H_q(B_1) + H_q(B_2) \to H_q(B) \to \cdots$$

$$\cdots \to H_q(B_0, A_0) \to H_q(B, A) \xrightarrow{j_1 - j_2} H_q(B_1, B_0 + A_1) + H_q(B_2, B_0 + A_2)$$
$$\xrightarrow{\partial_1 + \partial_2} H_{q-1}(B_0, A_0) \to \cdots$$

$$\cdots \to H_{q+1}(B, A) \xrightarrow{\partial_0} H_q(B_0, A_0) \to H_q(B_1, A_1) + H_q(B_2, A_2)$$
$$\to H_q(B, A) \to \cdots.$$

Here $\partial_1 : H_q(B_1, B_0 + A_1) \to H_{q-1}(B_0, A_0)$ is defined by the composite

$$\begin{array}{ccccc} H_q(B_1, B_0 + A_1) & \xrightarrow{\partial} & H_{q-1}(B_0 + A_1) & \to & H_{q-1}(B_0 + A_1, A_1) \\ & \searrow^{\partial_1} & & & \uparrow \cong \\ & & & & H_{q-1}(B_0, A_0) \end{array}$$

$j_1 : H_q(B, A) \to H_q(B_1, B_0 + A_1)$ is defined by the composite

$$\begin{array}{ccc} H_q(B, A) & \to & H_q(B, B_2 + A) \\ & \searrow^{j_1} & \uparrow \cong \\ & & H_q(B_1, B_0 + A_1) \end{array}$$

and so forth, so that $\partial_0 = \partial_1 j_1 = \partial_2 j_2$, etc.

I.3.2 Theorem. *(Sum Theorem for Poincaré pairs). With notation as above, any two of the following conditions imply the third:*

(i) *(B, A) is a Poincaré chain pair with orientation $v \in H_m(B, A)$*

(ii) *(B_0, A_0) is a Poincaré chain pair with orientation $\partial_0 v \in H_{m-1}(B_0, A_0)$*

(iii) *$(B_i, B_0 + A_i)$ are Poincaré chain pairs with orientations $v_i = j_i(v) \in H_m(B_i, B_0 + A_i)$, $i = 1, 2$.*

Proof. From (I.3.1) we have the following commutative diagram (up to sign) with the Mayer-Vietoris sequences:

$$\begin{array}{ccccccc} \to H^{q-1}(B_0) & \xrightarrow{\delta} & H^q(B) & \to & H^q(B_1) + H^q(B_2) & \to \cdots \\ \downarrow \partial_0 v \cap & & \downarrow v\cap & & \downarrow v_1 \cap + v_2 \cap & \\ \to H_{m-q}(B_0, A_0) & \to & H_{m-q}(B, A) & \to & H_{m-q}(B_1, B_0 + A_1) + H_{m-q}(B_2, B_0 + A_2) & \to \cdots \end{array}$$

The result then follows from the Five Lemma. □

I.3.3 Corollary. *Let (B, A), (B', A') be Poincaré pairs with $B = B_1 + B_2$, $B' = B'_1 + B'_2$, as above. Suppose $f: (B, A) \to (B', A')$ is a chain map $f(B_i) \subset B'_i$. Suppose (B_0, A_0) and (B'_0, A'_0) are Poincaré pairs with orientations $\partial_0 v$, $\partial_0 v'$, (v, v' are orientations of (B, A), (B', A') respectively). Then the following three conditions are equivalent:*

(a) *f has degree 1.*

(b) *$f_0 = f|(B_0, A_0)$ has degree 1 with respect to the orientation $\partial_0 v$.*

(c) *$f_i = f|(B_i, B_0 + A_i)$ have degree 1 with respect to the orientations v_i.*

Proof. Consider the maps of Mayer-Vietoris sequences induced by f, and the result is immediate. □

Thus (I.3.2) and (I.3.3) allow one to define the *sum* of Poincaré pairs and the *sum* of maps of degree 1, namely (B, A) is the *sum* of $(B_1, B_0 + A_1)$ and $(B_2, B_0 + A_2)$ along (B_0, A_0), and f is the *sum* of f_1 and f_2. Note that the orientations must be compatible.

Another refinement of Poincaré duality is the following:

I.3.4 Theorem. *Let (B, A) be a Poincaré chain pair of dimension m, and suppose $A = A_1 + A_2$, $A_0 = A_1 \cap A_2$, (A, A_i) are geometric chain pairs, $i = 0, 1, 2$, and A_0 is a Poincaré chain complex of dimension $m-2$, with orientation $\partial_0 \partial(v)$, where ∂v is the orientation of A, ∂_0 as above. Then $v \cap : H^q(B, A_1) \to H_{m-q}(B, A_2)$ is an isomorphism for all q.*

We call $(B; A_1, A_2)$ a Poincaré chain *triad*, and the analogous situation for spaces $(X; Y_1, Y_2)$ will be called a Poincaré *triad*.

Proof. We consider the diagram

$$\begin{array}{ccccccccc}
\cdots \to H^{q-1}(A_1, A_0) & \xrightarrow{\delta} & H^q(B, A) & \to & H^q(B, A_2) & \to & H^q(A_1, A_0) & \to \cdots \\
\downarrow {\scriptstyle v_1 \cap} & & \downarrow {\scriptstyle v \cap} & & \downarrow {\scriptstyle v \cap} & & \downarrow {\scriptstyle v_1 \cap} & \\
\cdots \to H_{m-q}(A_1) & \to & H_{m-q}(B) & \to & H_{m-q}(B, A_1) & \to & H_{m-q-1}(A_1) & \to \cdots .
\end{array}$$

Here δ is defined by the composite:

$$H^{q-1}(A_1, A_0) \xleftarrow{\cong} H^{q-1}(A, A_2) \xrightarrow{\delta'} H^q(B, A)$$

where δ' is the coboundary of the triple (B, A, A_2), so that the upper row is the exact sequence of this triple with $H^*(A, A_2)$ replaced by the isomorphic $H^*(A_1, A_0)$. Here $v_1 = $ image of $\partial'(v)$, where

$$\partial' : H_m(B, A) \to H_{m-1}(A, A_2),$$

v is the orientation of (B, A). It follows from (I.3.1) that the diagram is commutative up to sign, and from (I.3.2) it follows that $v_1 \cap$ is an isomorphism. Hence the result follows from the Five Lemma. □

All the results may be applied to topological spaces, where the chain complex of a pair of spaces (X, Y), $Y \subset X$, is a geometric chain pair, (where the diagonal map of the chain complex is induced by the diagonal map of spaces $x \mapsto (x, x)$, and the Eilenberg-Silber map, for singular chains [21]).

Our results yield a proof of the Poincaré duality theorem for differentiable manifolds. First let us recall the notions of orientability for manifolds.

Let M be an m-dimensional manifold with boundary ∂M, so that each point in M has a neighborhood homeomorphic to the closed unit ball D^m in R^m (as usual we assume M is Hausdorff). Points in $\operatorname{int} M = M - \partial M$ have neighborhoods homeomorphic to the open ball, while points in ∂M do not. It follows by excision that for $x \in \operatorname{int} M$,

$$H_m(M, M - x) = H_m(D^m, D^m - 0) = \mathbb{Z}.$$

By an *orientation* of M, we will mean a choice of generator γ_x of $H_m(M, M - x)$ for each $x \in \operatorname{int} M$ which are compatible in the following sense:

Let l be a simple differentiable arc in $\operatorname{int} M$ joining x_1 to x_2. Then the inclusions

$$\xi_i : M - l \to M - x_i, \quad i = 1, 2,$$

are homotopy equivalences, so

$$\xi_{i*} : H_m(M, M - l) \to H_m(M, M - x_i)$$

are isomorphisms, $i = 1, 2$, and

$$\xi_{2*} \xi_{1*}^{-1}(\gamma_{x_1}) = \gamma_{x_2}.$$

I.3.5 Theorem. *If $(M, \partial M)$ is a compact orientable differential m-manifold with boundary, then $(M, \partial M)$ is a Poincaré pair.*

Proof. Use induction on the dimension m, $m = 0$ being trivial, so we assume the theorem proved for dimension $m - 1$. We then have the following special case:

I.3.6 Lemma. *(D^m, S^{m-1}) is a Poincaré pair, where D^m is the unit disk in R^m.*

Proof. $H^i(D^m, S^{m-1}) = \mathbb{Z}$ for $i = m$, 0 otherwise, $H^i(D^m) = \mathbb{Z}$ for $i = 0$, 0 otherwise. Since cup product has a unit, $\psi : H^*(D^m) \otimes H^*(D^m, S^{m-1}) \to \mathbb{Z}$, where $\psi(x \otimes y) = \langle x \cup y, v \rangle$, ($v$ generates $H_m(D^m, S^{m-1})$) is non-singular.

Then it follows from (I.1.8) that (D^m, S^{m-1}) is a Poincaré pair with orientation v, and the lemma is proved.

Now we use results of M. Morse (as exposed in [41]). Let us assume that M is connected. By choosing a function f on M with one minimum in the interior of M and isolated non-degenerate critical points, we may use the Morse lemma to write

$$M=\bigcup_{i=0}^{q} M_i,\ M_i\subset M_{i+1},\ M_0=D_0^m,\ M_{i+1}=M_i\cup D_{i+1}^m\,,$$

and $M_i\cap D_{i+1}^m=N_i^{m-1}$, a compact differentiable manifold with boundary, $N_i=S^k\times D^{m-k-1}$ some $k\geqq 0$, $N_i^{m-1}\subset\partial M_i^m\cap\partial D_{i+1}^m$. Now use induction on i. For $i=0$, $(M_0,\partial M_0)=(D_0^m,S_0^{m-1})$ is a Poincaré pair by (I.3.6). Suppose $(M_i,\partial M_i)$ is a Poincaré pair. Then since $(N_i,\partial N_i)$ is a differentiable manifold with boundary, of dimension $m-1$, by induction (on m) it is a Poincaré pair. Therefore if the orientations are compatible, we may apply (I.3.2) to show $M_{i+1}=M_i\cup D_{i+1}^m$ and its boundary is a Poincaré pair.

Now $N_i=S^k\times D^{m-k-1}$, for some k, and if $k>0$, N_i is connected so that one can choose the orientations of M_i, D_{i+1}^m and N_i compatibly, since both orientations of N_i come from orientations of M_i and D_{i+1}^m. If $k=0$, however, there are four orientations of $S^0\times D^{m-1}$, and if we take one coming from ∂M_i, we must show it comes from an orientation of ∂D_{i+1}^m.

Now M^m is an orientable manifold, so that each $M_i^m\subset M^m$ is also orientable. That means that one can choose generators for $H_m(M,M-x)$, each $x\in\operatorname{int}M$, in a compatible way, so that if l is a simple curve joining x_0 to x_1, there is a generator of $H_m(M,M-l)$ which goes onto the given generators of $H_m(M,M-x_i)$, $i=0,1$.

I.3.7 Lemma. *If $(M,\partial M)$ is an n-manifold and an n-dimensional Poincaré pair, then the image*

$$j_{x*}[M]\in H_n(M,M-x),\quad j_x:(M,\partial M)\to(M,M-x)\,,$$

all $x\in\operatorname{int}M$, defines an orientation of M in the sense of manifolds, where $[M]\in H_n(M,\partial M)$ defines an orientation of $(M,\partial M)$ as a Poincaré pair.

To define the compatible orientation of ∂M at a point $y\in\partial M$, we take a curve $l\subset M$ such that the initial point is y and rest of the curve lies in $\operatorname{int}M$. Then the orientation of M defines a generator of $H_m(M,(M-l)\cup\partial M)$. Then the boundary operator for the triad $(M;M-l,\partial M)$ is a map

$$\partial:H_m(M,(M-l)\cup\partial M)\to H_{m-1}(\partial M,\partial M\cap(M-l))=H_{m-1}(\partial M,\partial M-y),$$

and we take the image of the generator of $H_m(M,(M-l)\cup\partial M)$ to define the compatible orientation of ∂M. One shows easily that it is well defined.

Now suppose again $N_i=M_i\cap D_{i+1}^m=\partial M_i\cap\partial D_{i+1}^m$, etc., $N=S^0\times D^{m-1}$ and M_i and D_{i+1} are oriented as Poincaré complexes so that the induced orientations (as manifolds) agree on $-1\times D^{m-1}\subset S^0\times D^{m-1}$. It follows that this choice determines an orientation on $M_{i+1}-(1\times D^{m-1})$. Similarly we may orient M_i and D_{i+1} so that the orientations are compatible on $1 \times D^{m-1}$, so determine an orientation on $M_{i+1} - (-1 \times D^{m-1})$, so that the two orientations agree on interior D^m_{i+1}. Since M_{i+1} is orientable, these orientations of $M_{i+1} - (-1 \times D^{m-1})$ and $M_{i+1} - (1 \times D^{m-1})$ come from an orientation of M_{i+1}, so that the orientations of M_i and D_{i+1} are compatible on $N = S^0 \times D^{m-1}$. □

A similar argument may be applied to piecewise linear manifolds using skeletons and regular neighborhoods instead of Morse functions.

§ 4. The Spivak Normal Fibre Space

In this section we introduce the Spivak normal fibre space of a Poincaré duality space (see [57]), This space will play an analogous role to that played by the normal bundle of a differentiable manifold in a high dimensional sphere. First we give some results which describe how to obtain spherical fibre spaces.

I.4.1 Theorem. *Let (X, Y) be a Poincaré duality pair of dimension $n+k$, X 1-connected, $\pi_2(X, Y) = 0$, and suppose X is a Poincaré duality space of dimension n. Then the inclusion map $i: Y \to X$ is equivalent to a fibre map with fibre the homotopy type of S^{k-1}.*

We also have a relative version:

I.4.2 Theorem. *Let (X, Y) be a Poincaré duality pair in dimension $n+k$, with X 1-connected. Let $Y = Y_1 \cup Y_2$, $Y_0 = Y_1 \cap Y_2$, $\pi_2(X, Y_1) = 0$, and suppose Y_0 is Poincaré duality space of dimension $n+k-2$ and (X, Y_2) is a Poincaré pair of dimension n. Then the inclusion map $i: Y_1 \to X$ is equivalent to a fibre map whose fibre is homotopy equivalent to S^{k-1}.*

I.4.3 Lemma. *Let $\pi: E \to B$ be a fibre map with 1-connected fibre F and base B. Then F is a homotopy S^{k-1} if and only if $H^i(\pi) = 0$ for $i < k$, and there is an element $U \in H^k(\pi)$ such that $\cup U: H^m(B) \to H^{m+k}(\pi)$ is an isomorphism for all m (i.e. the Thom isomorphism holds).*

Proof. Since F and B are 1-connected, so is E. Since π is a fibre map, $\pi_i(F) \to \pi_{i+1}(\pi)$ is an isomorphism for all i (see [28]).

If $F \cong S^{k-1}$ then $\pi_i(S^{k-1}) = 0$, $i < k-1$, so $\pi_i(\pi) = 0$ for $i < k$, and by the Relative Hurewicz Theorem $H_i(\pi) \cong \pi_i(\pi) = 0$ for $i < k$ and $H_k(\pi) \cong \pi_k(\pi) \cong \pi_{k-1}(F) = \mathbb{Z}$. Then by the Universal Coefficient Theorem $H^k(\pi) = \mathbb{Z}$, and let U be a generator. Then $\cup U: H^q(B) \to H^{q+k}(\pi)$ comes from a cochain map, namely $\cup u$, where u is a cochain representing U, and $\cup u: C^q(B) \to C^{q+k}(\pi)$. Since $i^* U$ is a generator of $H^k(cF, F)$, $i: (cF, F) \to (\bar{E}, E)$, $\bar{E}$ the associated cone fibre space of E, (see [7; Appendix] and (I.4.5) below) it follows that $\cup u$ preserves the filtrations in $C^q(B)$ and $C^{q+k}(\pi)$, and induces a map of the spectral sequences

$E_2(B) \to E_2(\pi)$. The spectral sequence $E_2(B)$ is trivial, $H^*(B) \cong E_2(B) \cong E_\infty(B)$, and $E_2^{p,q}(\pi) \cong H^p(B; H^q(cF, F))$ and $E_\infty(\pi) \cong G(H^*(\pi))$, ($G$ meaning the associated graded group). Since $F \cong S^{k-1}$, $H^q(cF, F) = 0$, $q \neq k$, $H^k(cF, F) = \mathbb{Z}$. It follows since d_r changes fibre degree for $r \geqq 2$, that $d_r \equiv 0$ all r, in $E_r(\pi)$, and $E_2(\pi) = E_\infty(\pi)$. Now $\cup U: H^p(B) \to H^p(B, H^k(cF, F))$ is an isomorphism, so the map $E_2(B) \to E_2(\pi)$ is an isomorphism. Hence $E_\infty(B) \to E_\infty(\pi)$ is an isomorphism, and since the associated graded groups are mapped isomorphically, we have $\cup U: H^p(B) \to H^{p+k}(\pi)$ is an isomorphism for all p.

Now let us suppose that $H^i(\pi) = 0$, $i < k$, and $U \in H^k(\pi)$ such that $\cup U: H^q(B) \to H^{q+k}(\pi)$ is an isomorphism for all q. Then by the Relative Hurewicz Theorem $H_i(\pi) = \pi_i(\pi) = 0$ for $i < k$, so $\pi_i(\pi) = \pi_{i-1}(F) = 0$ for $i < k$. Now $H^k(\pi) \cong H^0(B) = \mathbb{Z}$, since B is 0-connected, while

$$H^{k+1}(\pi) \cong H^1(B) = 0\,,$$

and hence $H_k(\pi) = \mathbb{Z}$, by the Universal Coefficient Theorem, and $H_k(\pi) \cong \pi_k(\pi) \cong \pi_{k-1}(F) = \mathbb{Z}$. Since F is 1-connected, it remains to show that $H^{i-1}(F) \cong H^i(cF, F) = 0$ for $i \neq k$, in order that F should be homotopy equivalent to S^{k-1}.

Now since $\cup U: H^q(B) \to H^{q+k}(\pi)$ is an isomorphism, and since $E_2^{p,q}(\pi) \cong H^p(B; H^q(cF, F))$ it follows first that

$$H^k(\pi) \cong H^0(B; H^k(cF, F)) \cong H^0(B) = \mathbb{Z}\,,$$

U is a generator, and that

$$\cup U: E_2^{p,0}(B) \to E_2^{p,k}(\pi) \cong H^p(B; H^k(cF, F))$$

is an isomorphism for all p.

If $\cup U: E_\infty(B) \to E_\infty(\pi)$ is not a monomorphism, then for some $x \in F^i H^*(B)$, $x \cup U \in F^{i+l} H^*(\pi)$ for some $l \geqq 1$. Then $x \cup U$ represents an element in $E_\infty^{i+l,k-l}(\pi)$ for some $l \geqq 1$. Since $E_2^{p,q}(\pi) = H^p(B; H^q(cF, F)) = 0$ for $q < k$, it follows that $E_\infty^{i+l,k-l}(\pi) = 0$ for $l \geqq 1$, and hence $x \cup U = 0$. Since $\cup U: H^*(B) \to H^*(\pi)$ is an isomorphism, it follows $x = 0$ and hence $\cup U: E_\infty^{p,0}(B) \to E_\infty^{p,k}(\pi)$ is a monomorphism.

Since $E_2^{p,q}(\pi) = 0$ for $q < k$, it follows that $E_\infty^{p,k}(\pi)$ is a quotient of $E_2^{p,k}(\pi)$, and since $\cup U: E_2^{p,0}(B) \to E_2^{p,k}(\pi)$ is an isomorphism and $\cup U: E_\infty^{p,0}(B) \to E_\infty^{p,k}(\pi)$ is a monomorphism, it follows that

$$\cup U: E_\infty^{p,0}(B) \to E_\infty^{p,k}(\pi)$$

is an isomorphism and $E_2^{p,k}(\pi) = E_\infty^{p,k}(\pi)$ for all p, and hence

$$E_r^{p,k}(\pi) \cap \text{image } d_r = 0$$

for all p, $r \geqq 2$.

Suppose $H^j(cF, F)=0$ for $k<j<l$, and $H^l(cF, F)\neq 0$. Then

$$E_2^{0,l}(\pi)\cong H^0(B; H^l(cF, F))\cong H^l(cF, F)$$

and the only group $E_2^{p,q}(\pi)$ with $q<l$ is $E_2^{p,k}(\pi)$. Since $E_2^{p,k}(\pi)\cap \text{image } d_r=0$, all r, p, it follows that $E_2^{0,l}(\pi)=E_\infty^{0,l}(\pi)$, and hence $H^l(\pi)$ contains some element x, such that $i^*(x)\neq 0$ in $H^l(cF, F)$, $l>k$. But, if $y\in H^*(B)$ then

$$i^*(y\cup U)=j^*(y)\cup i^*(U), \quad j:(\text{point})\to B.$$

Hence $i^*(y\cup U)=0$ if $y\in H^q(B)$, $q>0$, and since $\cup U$ is an isomorphism, it follows that $i^*(H^l(\pi))=i^*(H^{l-k}(B)\cup U)=0$ if $l>k$. It follows that $H^j(cF, F)=0$ for $j\neq k$, hence F is a homology S^{k-1}, 1-connected, so by the Theorem of J. H. C. Whitehead, $F\cong S^{k-1}$. □

Proof of (I.4.1). Replace the inclusion $i: Y\to X$ by a fibre map $\pi: E\to X$, with fibre F. Then $\pi_1(F)\cong\pi_2(X, Y)=0$, and we may apply (I.4.3) as follows: Let $\mu\in H_n(X)$ be such that $\mu\cap: H^q(X)\to H_{n-q}(X)$ is an isomorphism for all q, since X is an n-dimensional Poincaré duality space. Let $v\in H_{n+k}(X, Y)$ be a generator, so that $v\cap: H^q(X, Y)\to H_{n+k-q}(X)$ is an isomorphism for all q, since (X, Y) is an $(n+k)$-dimensional Poincaré pair. Let $U\in H^k(X, Y)$ be such that $v\cap U=\mu$. Also note that $H^i(X)=0$ for $i>n$, so $H_j(X, Y)=0$ for $j<k$. If $x\in H^q(X)$, by (I.1.2)

$$(v\cap(x\cup U))=(v\cap U)\cap x=\mu\cap x.$$

Hence $(v\cap)\circ(\cup U)=\mu\cap$ is an isomorphism, and $(v\cap)$ is an isomorphism, so $\cup U: H^q(X)\to H^{q+k}(X, Y)$ is an isomorphism. But

$$H^*(\pi)=H^*(i)=H^*(X, Y),$$

and applying (I.4.3) it follows that the fibre of π is homotopy equivalent to S^{k-1}. □

Proof of (I.4.2). Replacing $i: Y_1\to X$ by a fibre map $\pi: E\to X$ with fibre F we note that $\pi_1(F)=\pi_2(X, Y_1)=0$.

Now let us apply Theorem (I.3.4) which tells us that $(X; Y_1, Y_2)$ is a Poincaré triad, and $v\cap: H^q(X, Y_i)\to H_{n+k-q}(X, Y_j)$ is an isomorphism, $(i,j)=(1,2)$ or $(2,1)$, v a generator of $H_{n+k}(X, Y)$. Since (X, Y_2) is an n-dimensional Poincaré pair, $H^j(X, Y_2)=0$ for $j>n$, and $\mu\in H_n(X, Y_2)$ such that $\mu\cap: H^q(X, Y_2)\to H_{n-q}(X)$ is an isomorphism. Then

$$H^q(X, Y_2)\cong H_{n+k-q}(X, Y_1)$$

so $H_i(X, Y_1)=0$ for $i<k$. Let $U\in H^k(X, Y_1)$ such that $v\cap U=\mu$. Then if $x\in H^q(X)$, by (I.1.2), $v\cap(x\cup U)=(v\cap U)\cap x=\mu\cap x$, so $(v\cap)\circ(\cup U)=\mu\cap$ is an isomorphism and $v\cap$ is an isomorphism so $\cup U$ is an isomorphism. Hence by (I.4.3), the fibre map π equivalent to $i: Y_1\to X$ has fibre a homotopy S^{k-1}. □

For a spherical fibre space ξ, we define its Thom complex

$$T(\xi) = B \bigcup_{\pi} (cE_0),$$

where $\pi: E_0 \to B$ is the projection of the total space E_0 onto the base B, cE_0 is the cone on E_0.

I.4.4 Theorem. (Spivak). *Let (X, Y) be an n-dimensional Poincaré duality pair, with X 1-connected, Y a finite complex, up to homotopy type. Then there is a spherical fibre space $(\xi): \pi: E_0 \to X$ with fibre a homotopy S^{k-1}, and an element $\alpha \in \pi_{n+k}(T(\xi), T(\xi \mid Y))$ such that*

$$h(\alpha) \cap U = [X] \in H_n(X, Y).$$

We call ξ the *Spivak normal fibre space* of (X, Y).

Proof of (I.4.4). Let us suppose first $Y = \emptyset$, so X is a Poincaré duality space. Since X is 1-connected and $H^j(X) = 0$ for $j > n$ it follows easily that there is a finite n-complex K and a homotopy equivalence $g: X \to K$ (see [16, Exposé 22 Appendix]). Using standard arguments $K \subset S^{n+k}$ as a subpolyhedron for sufficiently large k, $k \geqq 3$. Let U^{n+k} be a regular neighborhood of K in S^{n+k}, so that U^{n+k} is an $(n+k)$-manifold with boundary ∂U^{n+k}, $U \subset S^{n+k}$, and $g': X \to U$ is a homotopy equivalence, ($g' = (\text{inclusion}) \circ g$). Since K is an n-complex, U an $(n+k)$-manifold, it follows from general position arguments that $\pi_i(U - K) \to \pi_i(U)$ is an isomorphism for $i < k-1$ and onto for $i = k-1$. Since U is a regular neighborhood of K, $\partial U \to (U - K)$ is a homotopy equivalence, so $\pi_i(\partial U) \to \pi_i(U)$ is an isomorphism for $i < k-1$, onto for $i = k-1$. Hence $\pi_i(U, \partial U) = 0$ for $i \leqq k-1$. Since X is 1-connected, U is 1-connected and since $k \geqq 3$ ∂U is 1-connected and $\pi_2(U, \partial U) = 0$. Then $(U, \partial U)$ satisfied the conditions of (I.4.1), so that the inclusion $\partial U \to U$ is equivalent to a fibre map with fibre a homotopy S^{k-1}. The pull back of this fibre space to X is $(\xi): \pi: E_0 \to X$.

Then $T(\xi) = X \bigcup_{\pi} (cE_0)$, so it follows that $T(\xi) \cong U \bigcup_{i} (c\partial U) \cong U/\partial U$. Since $U^{n+k} \subset S^{n+k}$, the natural collapse $\eta: S^{n+k} \to U/\partial U$ has the property that $\eta_*(\text{generator}) = \text{generator of } H_{n+k}(U/\partial U)$. Hence the homotopy class α of the corresponding map $S^{n+k} \to T(\xi)$ has the required properties.

In case $Y \neq \emptyset$, we will make a similar argument using (I.4.2) instead of (I.4.1). One embeds Y in S^{n+k-1} using the fact that it is the homotopy type of a finite complex. As above, X is the homotopy type of a finite complex, and we may assume (replacing X by the mapping cylinder of $Y \to X$) that Y is a subcomplex of X with a neighborhood $Y \times I \subset X$, $Y = Y \times 0 \subsetneqq Y \times I$. Extending to an embedding of $Y \times I \to D^{n+k}$ such that the coordinate in I goes into the radial coordinate in D^{n+k}, we get $Y \times 1 \subset \text{Interior } D^{n+k}$, and if k is very large we may extend to an embedding of X in D^{n+k} with $X \cap S^{n+k-1} = Y$. Then one may apply (I.4.2) to the regular neighborhood of X in D^{n+k}, where the intersection of this neighborhood with S^{n+k-1} is a regular neighborhood of Y. (One uses the star of X in the second derived subdivision.)

Let B = regular neighborhood of X in D^{n+k}, $C = B \cap S^{n+k-1}$ = regular neighborhood of Y in S^{n+k-1}, E = closure of $\partial B - C$, $E_1 = E \cap C = \partial E = \partial C$. Applying Theorem (I.4.2), it follows that $E \to B$ is equivalent to a spherical fibre space ξ, (which we denote by the same letters). However, $E_1 \to C$ may not be spherical. Set $\xi' = i^*(\xi): E' \to C$, the induced spherical fibre space over C. Since the diagram

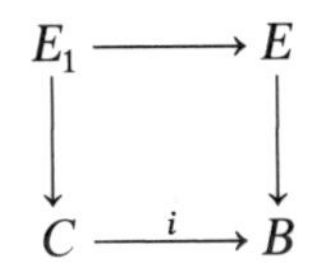

commutes, we may factor $E_1 \to E$ through E', so that we have a map of pairs $e:(E, E_1) \to (E, E')$ lying over the identity map of (B, C). Since (E, E') is a fibred pair over (B, C) with fibre a homotopy S^{k-1} and (B, C) is a Poincaré pair (it is homotopy equivalent to (X, Y)) it follows that (E, E') is a Poincaré pair and that $(E, E_1) \to (E, E')$ is a map of degree 1. Since $e: E \to E$ is the identity, it follows that $e_*: H_*(E, E_1) \to H_*(E, E')$ is an isomorphism, using (I.2.7). Hence we get $\bar{e}: (B/E, C/E_1) \to (B/E, C/E')$, and $B/E = T(\xi)$, $B/E' = T(\xi')$, so $\bar{e}: (B/E, C/E_1) \to (T(\xi), T(\xi'))$, and $\bar{e}_*$ is an isomorphism in homology. There is the natural collapsing map $\alpha: (D^{n+k}, S^{n+k-1}) \to (B/E, C/E_1)$ so that $\bar{e}_*(\alpha) \in \pi_{n+k}(T(\xi), T(\xi'))$ has the property $h(\bar{e}_*(\alpha)) \cap U = [X] \in H_n(B, C) \cong H_n(X, Y)$, which proves (I.4.4) for $Y \neq \emptyset$. □

Before we go on to study the properties of the Spivak normal fibre space we first recall some properties of spherical fibre spaces.

I.4.5. *Any spherical fibre space can be embedded (up to fibre homotopy equivalence) as a subfibre space of a fibre space with contractible fibre (analogous to the disk bundle for a linear fibre space).*

This may be proved by first replacing the projection by an inclusion, then replacing the inclusion by the space of paths fibration, so that the contractible fibre is the path space of the base. We leave details to the reader (cf. [7, Appendix]).

For a spherical fibre space ξ we will denote its total space by $E_0(\xi)$, and $E_0(\xi) \subset E(\xi)$ = the fibre space with contractible fibre. With the aid of $E(\xi)$ we may now imitate some of the contructions of linear bundle theory in the category of spherical fibre spaces. For example to define Whitney sum of ξ_1 and ξ_2, we first take $E(\xi_1) \times E(\xi_2)$ over $X \times X$ and define $E(\xi_1 + \xi_2) = \Delta^*(E(\xi_1) \times E(\xi_2))$ $\Delta: X \to X \times X$, the diagonal. Then $E_0(\xi_1 + \xi_2) = \Delta^*(E_0(\xi_1) \times E(\xi_2) \cup E(\xi_1) \times E_0(\xi_2))$.

It is easy to see that the Thom complex $T(\xi) \cong E_0(\xi+\varepsilon^1)/\alpha X$, where ε^1 is the trivial line bundle, $\alpha: X \to E_0(\xi+\varepsilon^1)$ is the canonical cross section.

If α is a spherical fibre space let $\mathrm{End}(\alpha)$ denote the group of fibre homotopy classes of fibre maps $\alpha \to \alpha$ covering the identity of the base space. Then there are natural maps $\mathrm{End}(\alpha) \to \mathrm{End}(\alpha+\varepsilon^1)$, $f \mapsto f+1$. Define the stable equivalences of α to be $\mathscr{E}(\alpha) = \lim_{n\to\infty} \mathrm{End}(\alpha+\varepsilon^n)$. Clearly $\mathscr{E}(\alpha) = \mathscr{E}(\alpha+\varepsilon^1)$. Now $\mathrm{End}(\alpha) \to \mathrm{End}(\alpha+\alpha^{-1}) = \mathrm{End}(\varepsilon^{2q})$, $f \mapsto f+1$, defines a map $a: \mathscr{E}(\alpha) \to \mathscr{E}(\varepsilon)$ and $\mathrm{End}(\varepsilon^q) \to \mathrm{End}(\alpha+\varepsilon^q)$, $g \mapsto 1+g$ defines $b: \mathscr{E}(\varepsilon) \to \mathscr{E}(\alpha)$, and clearly $ab=1$, $ba=1$. So we get

I.4.6 Lemma. *The group of stable equivalences is independent of the fibre space, i.e. $\mathscr{E}(\alpha) \cong \mathscr{E}(\beta)$, any two spherical fibre spaces α, β over X.*

Clearly the result above holds for any category of fibre spaces, such as linear bundles, piecewise linear bundles, or topological fibre bundles with R^n as fibre, as long as Whitney sum and inverses are defined, and was first proved by Hirsch and Mazur [29].

I.4.7 Proposition. $\mathrm{End}(\alpha) = [X, G_q]$, *if α^q is fibre homotopically trivial, where G_q = space of homotopy equivalences of S^{q-1} to itself, with the compact-open topology.*

Proof. Consider first the product space $X \times S^{q-1}$. Pick a point $x \in X$, and consider $j_x: S^{q-1} \to X \times S^{q-1}$, $j_x(s) = (x, s)$, $s \in S^{q-1}$, and the projection $p: X \times S^{q-1} \to S^{q-1}$. If $f \in \mathrm{End}(\varepsilon^q)$, then define $\eta(f): X \to G_q$ by

$$\eta(f)(x) = p f j_x : S^{q-1} \to S^{q-1}.$$

Since f is a fibre homotopy equivalence it follows that $\eta(f)(x)$ is a homotopy equivalence, so $\eta(f)(x) \in G_q$. It is easy to verify that $\eta(f)$ is a continuous map and that a fibre homotopy is sent into a homotopy of $\eta(f)$, so that $\eta: \mathrm{End}(\varepsilon^q) \to [X, G_q]$.

If $\alpha: X \to G_q$ then α defines a continuous map $\varrho(\alpha): X \times S^{q-1} \to S^{q-1}$, using the "exponential" law $((S^{q-1})^{S^{q-1}})^X = (S^{q-1})^{X \times S^{q-1}}$. Define $\gamma(\varrho(\alpha)): X \times S^{q-1} \to X \times S^{q-1}$ by $\gamma(\varrho(\alpha))(x, s) = (x, \varrho(\alpha)(x, s))$. It follows that $\gamma(\varrho(\alpha))$ is a map of fibre spaces, and a homotopy equivalence on each fibre. One checks easily that $\gamma\varrho$ defines a map $\omega: [X, G_q] \to \mathrm{End}(\varepsilon^q)$ and $\eta\omega = 1$, $\omega\eta = 1$, so the groups are isomorphic. The proof is now completed by:

I.4.8 Lemma. *If $b: \alpha \to \beta$ is a fibre homotopy equivalence, then b induces an isomorphism $b_*: \mathrm{End}(\alpha) \to \mathrm{End}(\beta)$.*

Proof. Let $b': \beta \to \alpha$ be an inverse for b so that bb' and $b'b$ are fibre homotopic to the identity. If $f: \alpha \to \alpha$, define $b_\# f: \beta \to \beta$ by $b_\# f = b' f b$. It is easy to verify this induces an isomorphism $b_*: \mathrm{End}(\alpha) \to \mathrm{End}(\beta)$.

I.4.9 Proposition. *Let F_q = space of base point preserving homotopy equivalences $S^q \to S^q$, $F_q \subset G_{q+1}$. Then the evaluation $e: G_{q+1} \to S^q$, $e(f) = f(*)$, $* \in S^q$ a base point, defines a fibre map with fibre F_q.*

Proof. See Spanier [55].

Now F_q = the identity component of $\Omega^q S^q$, (see [55]) and the suspension of maps yields maps $G_{q+1} \to G_{q+2}$, $F_q \to F_{q+1}$, and the induced $\Omega^q S^q \to \Omega^{q+1} S^{q+1}$ is simply the usual suspension map. Hence by the Freudenthal Suspension Theorem $\pi_n(F_{q+1}, F_q) = 0$ for $n \leqq q-1$ (see [55]), and $[K, F_q] \cong [K, F_{q+1}]$ if the dimension of $K < q-1$. It follows from (I.4.9) that the same result holds for G_{q+1} since $F_q \subset G_{q+1}$ induces isomorphism on π_i $i < q-1$:

I.4.10 Proposition. $\pi_n(G_{q+1}, G_q) = 0$ *for* $n \leqq q-2$, *so* $[K, G_q] \cong [K, G_{q+1}]$ *for dimension* $K < q-2$.

We get from (I.4.7) and (I.4.10):

I.4.11 Corollary. $\mathrm{End}(\alpha^q) \cong \mathscr{E}(\alpha)$ *if* α *is fibre homotopy trivial and dimension of base space* $< q-2$.

Now we can prove:

I.4.12 Theorem. $\mathrm{End}(\alpha^q) \cong \mathrm{End}(\alpha^q + \varepsilon^1)$ *so* $\mathrm{End}(\alpha^q) = \mathscr{E}(\alpha)$, *provided the dimension of the base space* $< q-2$.

Proof. We proceed by induction on dimension and on the number of cells. If there is only one cell then α is fibre homotopy trivial since the base is contractible, and then the result follows from (I.4.11).

Now suppose $X = X_0 \cup e^n$, $n < q-2$, and $\mathrm{End}(\alpha | X_0) \cong \mathrm{End}(\alpha + \varepsilon^1 | X_0)$. Let $f: \alpha + \varepsilon^1 \to \alpha + \varepsilon^1$ be a fibre homotopy equivalence. Since $f|(\alpha + \varepsilon^1 | X_0)$ is homotopic to $g_0 + 1$, where $g_0: \alpha | X_0 \to \alpha | X_0$, using the covering homotopy theorem, we may assume (by changing f by a fibre homotopy) that $f|((\alpha + \varepsilon^1)| X_0) = g_0 + 1$.

Let $\partial e^n = S^{n-1} = e^n \cap X_0$, then $\alpha | e^n$ is fibre homotopy trivial, since e^n is contractible, and thus $\alpha | S^{n-1}$ is fibre homotopy trivial. Pick a fibre homotopy equivalence between $\alpha | e^n$ and $e^n \times S^{q-1}$, and keep it fixed during the remainder of the argument.

With this representation of $\alpha | e^n$, we get a representation of $f|(\alpha + \varepsilon^1 | e^n)$, as a map $f: e^n \times S^q \to e^n \times S^q$, degree 1 on each fibre, and $f | S^{n-1} \times S^q = S g_0$ where $g_0: S^{n-1} \times S^{q-1} \to S^{n-1} \times S^{q-1}$ and S means suspension on each fibre $x \times S^{q-1}$.

Hence f, g_0 define a map $a: (e^n, S^{n-1}) \to (G_{q+1}, G_q)$. But $\pi_n(G_{q+1}, G_q) = 0$ for $n < q-1$ by (I.4.10) so a is nullhomotopic. It follows that g_0 extends to $g: e^n \to G_q$ and f is fibre homotopic to $g+1$ keeping $f | S^{n-1}$ fixed. Hence $\mathrm{End}(\alpha) \to \mathrm{End}(\alpha + \varepsilon^1)$ is surjective. A similar argument about a fibre homotopy between $g_0 + 1$ and $g_1 + 1$ shows the map is injective. □

Now we recall the theory of S-duality as developed in Spanier [56]. If A and B are two spaces with base points $a_0 \in A$, $b_0 \in B$, the "wedge" $A \vee B = A \times b_0 \cup a_0 \times B \subset A \times B$. We denote the "smash" of A and B by $A \wedge B = A \times B / A \vee B$.

A map $\alpha: A \wedge B \to S^n$ will be called an n-duality map if

$$\alpha^*(g)/: H_q(A) \to H^{n-q}(B)$$

is an isomorphism for all q, where $g \in H^n(S^n)$ is a generator. A and B will be said to be n-dual in S-theory if some suspension $\Sigma^k A$ is homotopy equivalent to $S^{n+1+k+q} - \Sigma^q B$ for an embedding of $\Sigma^q B$ in $S^{n+1+k+q}$, k and q arbitrarily large.

Theorem (Spanier). *A and B are n-dual in S-theory if and only if there is a $n+k$-duality map $\Sigma^k A \wedge B \to S^{n+k}$ for some k.*

In the theory of S-duality developed by Spanier and Whitehead they consider the S-groups $\{X, Y\} = \varinjlim [\Sigma^k X, \Sigma^k Y]$, where $[A, B]$ = the set of homotopy classes of (base point preserving) maps of A to B. The equivalence class of $f: X \to Y$ in $\{X, Y\}$ is denoted by $\{f\}$. If A and A' are n-dual in S-theory and B and B' are n-dual in S-theory, then they defined $D_n: \{A, B\} \to \{B', A'\}$ which they proved to be an isomorphism of groups. If $f: A \to B$ is an inclusion, $B \subset S^{n+1}$, then clearly $S^{n+1} - B$ is included in $S^{n+1} - A$ and this inclusion represents $D_n(\{f\})$ in $\{B', A'\}$. The general case can be reduced to this by replacing B by a regular neighborhood in a high dimensional sphere.

In terms of n-duality maps Spanier [56] showed the following:

Theorem (Spanier). *Let $\alpha: A \wedge A' \to S^n$ and $\beta: B \wedge B' \to S^n$ be n-duality maps and let $f: A \to B$, $g: B' \to A'$. Then f and g are n-dual in S-theory ($\{g\} = D_n(\{f\})$) if and only if the following diagram (or some suspension of it) commutes up to homotopy*

$$\begin{array}{ccc} A \wedge B' & \xrightarrow{f \wedge 1} & B \wedge B' \\ \big\downarrow{\scriptstyle 1 \wedge g} & & \big\downarrow{\scriptstyle \beta} \\ A \wedge A' & \xrightarrow{\alpha} & S^n \end{array}.$$

Now if A and A' are n-dual and B and B' are m-dual, it follows easily that $A \wedge B$ and $A' \wedge B'$ are $(n+m)$-dual, (all in S-theory). Hence the condition $\alpha: A \wedge B \to S^k$ such that $\alpha^*(g)/: H_s(A) \to H^{k-s}(B)$ is an isomorphism all s is equivalent using S-duality and Alexander duality between homology and cohomology, with the following: $\beta: S^{m+n-k} \to A' \wedge B'$ such that $\beta_*(g)/: H^q(A') \to H_{m+n-k-q}(B')$ is an isomorphism all q. But taking $B = A'$, $B' = A$, $n = m = k$, we get $\beta: S^n \to B \wedge A$, $\beta_*(g)/: H^q(B) \to H_{n-q}(A)$, an isomorphism, all q. Thus we get an equivalent formulation of the two theorems of Spanier above:

An n-duality map is a map $\beta: S^n \to A \wedge B$ such that $\beta_*(g)/: H^q(A) \to H_{n-q}(B)$ is an isomorphism for all q.

I.4.13 Theorem. *A and B are n-dual in S-theory if and only if there exists an n-duality map $\beta: S^n \to A \wedge B$.*

I.4.14 Theorem. *Let $\alpha: S^n \to A \wedge A'$, $\beta: S^n \to B \wedge B'$ be n-duality maps, and let $f: A \to B$, $g: B' \to A'$. Then $\{g\} = D_n(\{f\})$ if and only if the diagram*

$$\begin{array}{ccc} S^n & \xrightarrow{\alpha} & A \wedge A' \\ \Big\downarrow{\scriptstyle \beta} & & \Big\downarrow{\scriptstyle f \wedge 1} \\ B \wedge B' & \xrightarrow{1 \wedge g} & B \wedge A' \end{array}$$

commutes up to homotopy.

Now we may prove (following Wall [67]) an enriched version of the uniqueness theorem for the Spivak normal fibre space.

We use this strengthened version of Atiyah's generalization [4] of the Milnor-Spanier Theorem [43]:

I.4.15 Theorem. *Let (X, Y) be a Poincaré duality pair of dimension m, X 1-connected, ν its Spivak normal fibre space as defined above. If ξ is another spherical fibre space over X then $T(\xi)/T(\xi \mid Y)$ is S-dual to $T(\nu + (-\xi))$ (where $-\xi$ is the inverse of ξ).*

Proof. We construct a duality map as in (I.4.13). Now

$$(\nu + (-\xi)) + \xi = \nu + \varepsilon, \quad \varepsilon = \text{trivial},$$

and $\nu + \varepsilon$ is then induced by the diagonal $\Delta: X \to X \times X$ from the fibre space $(\nu + (-\xi)) \times \xi$ over $X \times X$. Call $\varrho: (\nu + \varepsilon) \to (\nu + (-\xi)) \times \xi$ the map of fibre spaces. Consider the diagonal as a map of pairs $(X, Y) \to X \times (X, Y)$ and consider ϱ as a map of pairs

$$\begin{aligned} \bar{\varrho}: &(E(\nu + \varepsilon), E((\nu + \varepsilon) \mid Y) \cup E_0(\nu + \varepsilon)) \\ &\to (E(\nu + (-\xi)), E_0(\nu + (-\xi))) \times (E(\xi), E(\xi \mid Y) \cup E_0(\xi)). \end{aligned}$$

The subspace of the product pair is

$$(E_0(\nu + (-\xi)) \times E(\xi)) \cup (E(\nu + (-\xi)) \times (E(\xi \mid Y) \cup E_0(\xi)))$$

so that includes all of $E_0((\nu + (-\xi)) \times \xi) \cup E(((\nu + (-\xi)) \times \xi) \mid Y)$ so that $\bar{\varrho}$ is a map of pairs. Collapsing subspaces, $\bar{\varrho}$ induces a map of Thom complexes

$$\varrho': (T(\nu + \varepsilon), T(\nu + \varepsilon \mid Y)) \to (T(\nu + (-\xi)), \infty) \times (T(\xi), T(\xi \mid Y)).$$

Then the following diagram is commutative

$$\begin{array}{ccc} H_*(T(\nu+\varepsilon), T(\nu+\varepsilon|Y)) & \xrightarrow{\varrho'_*} & H_*((T(\nu+(-\xi), \infty)\times(T(\xi), T(\xi|Y))) \\ \downarrow{\scriptstyle \cap U_0} & & \downarrow{\scriptstyle \cap(U_1\cup U_2)} \\ H_*(X, Y) & \xrightarrow{\Delta_*} & H_*(X\times(X, Y)) \end{array} \qquad (*)$$

where U_i, $i=0, 1, 2$ are the three Thom classes, and $\varrho'^*(U_1\cup U_2)=U_0$. Since $\nu+\varepsilon$ is the Spivak normal fibre space of (X, Y), there is an $\alpha\in\pi_{m+k}(T(\nu+\varepsilon), T(\nu+\varepsilon|Y))$ such that $h(\alpha)\cap U_0=[X]\in H_m(X, Y)$. We claim

$$\varrho'\alpha : S^{m+k}\to T(\nu+(-\xi))\wedge(T(\xi)/T(\xi|Y))$$

is a duality map (see (I.4.13)). For any element in $H^*(T(\nu+(-\xi)))$ is of the form $x\cup U_1, x\in H^*(X)$, by the Thom isomorphism theorem. Then

$$\begin{aligned} ((\varrho'\alpha)_*(g)/(x\cup U_1))\cap U_2 &= (\varrho'_*(h(\alpha))/x\cup U_1\cup U_2) \\ &= (\varrho'_*(h(\alpha))\cap(U_1\cup U_2))/x = \Delta_*(h(\alpha)\cap U_0)/x \\ &= \Delta_*([X])/x = [X]\cap x\,. \end{aligned}$$

Thus since $[X]\cap$ is an isomorphism, and $\cap U_2$ and $\cup U_1$ are isomorphisms, it follows that $(\varrho'\alpha)_*(g)/$ is an isomorphism, and hence $\varrho'\alpha$ is a duality map. □

Now we wish to consider the relation between the isomorphism of (I.4.6) between the stable equivalences $\mathscr{E}(\xi)$ and $\mathscr{E}(\nu+(-\xi))$ and the duality (I.4.15) between the Thom complexes. We recall that if θ is a trivial fibre space and $b:\theta\to\theta$ is an equivalence of it, $b\mapsto b+1$ defines a homomorphism $\mathscr{E}(\theta)\to\mathscr{E}(\theta+\xi)$, and $b\mapsto 1+b$ defines

$$\mathscr{E}(\theta)\to\mathscr{E}(\nu+(-\xi)+\theta)$$

which induce the isomorphisms $\gamma:\mathscr{E}(\theta)\to\mathscr{E}(\xi)$ and $\gamma'\mathscr{E}(\theta)\to\mathscr{E}(\nu+(-\xi))$ of the stable equivalences (see (I.4.6), (I.4.12), etc.).

I.4.16 Theorem. *Using the duality of* (I.4.15) *between* $T(\xi)/T(\xi|Y)$ *and* $T(\nu+(-\xi))$, *then* $T(\gamma(b))$ *is dual to* $T(\gamma'(b))$.

Proof. We recall that the duality map of (I.4.15) is induced by the fibre space map $\nu+\varepsilon\to(\nu+(-\xi))\times\xi$ covering the diagonal considered as a map $X\to X\times(X, Y)$. Then the natural map $\alpha:S^{m+k}\to T(\nu+\varepsilon)/T(\nu+\varepsilon|Y)$ composed with the map induced by the fibre space map yields the duality using (I.4.13).

If we add two trivial factors θ we get

$$\varrho:\nu+\varepsilon'\to(\nu+(-\xi)+\theta)\times(\theta+\xi)\,.$$

On $((\nu+(-\xi))+\theta)\times(\theta+\xi)$ we may consider $b_1=(1+b)\times(1)$ and $b_2=(1)\times(b\times 1)$. Now $\nu+\varepsilon'=\nu+(-\xi)+\theta+\theta+\xi$, and on $\theta+\theta$ the equivalences $b+1$ and $1+b$ are homotopic. Then $b_1\varrho=\varrho(1+(b+1)+1)$ and $b_2\varrho=\varrho(1+(1+b)+1)$ on $(\nu+(-\xi))+(\theta+\theta)+\xi$ so $b_1\varrho$ is homotopic to $b_2\varrho$ as an equivalence of fibre spaces. It follows that $b_1\varrho'\alpha$ is homotopic to $b_2\varrho'\alpha$ and thus the diagram below commutes up to homotopy:

$$\begin{array}{ccc} S^{m+k} & \xrightarrow{\varrho'\alpha} & A\wedge B \\ {\scriptstyle \varrho'\alpha}\downarrow & & \downarrow{\scriptstyle 1\wedge T(b+1)} \\ A\wedge B & \xrightarrow{T(1+b)\wedge 1} & A\wedge B \end{array}$$

where $A=T(\nu+(-\xi)+\theta)$, $B=T(\theta+\xi)/T((\theta+\xi)|Y)$, ϱ' is as in (I.4.15) and $T(\ )$ indicates the induced map of Thom complexes. Then (I.4.14) implies that $T(b+1)$ on B is dual to $T(1+b)$ on A. □

Let θ be the trivial spherical fibre space of fibre dimension $k-1$ over B, $k>>\dim B$. Let $b:\theta\to\theta$ be an oriented fibre homotopy equivalence. and let $\beta\in\pi^k(T(\theta))$ be induced by a fixed fibre homotopy trivialization, $E^0(\theta)\to S^{k-1}$. Then if $h':\pi^k\to H^k$ is the Hopf homomorphism,

$$h'(g)=g^*(\text{generator}),\quad g:X\to S^k,$$

then $h'(\beta)$ is a Thom class of $T(\theta)$ since $j^*h'(\beta)=$ generator of $H^k(S^k)$, $j:S^k\to T(\theta)$ coming from the inclusion of the fibre, and $h'(T(b)^*(\beta))$ is also a Thom class. Let $\mathscr{E}_0(\xi)=$ the group of stable orientation preserving fibre homotopy equivalences of the fibre space ξ, $\mathscr{E}_0(\xi)\subset\mathscr{E}(\xi)$.

I.4.17 Proposition. *The map* $\psi:\mathscr{E}_0(\theta)\to\pi^k(T(\theta))$ *induced by* $\psi(b)=T(b)^*(\beta)$ *induces a* $1-1$ *correspondence between* $\mathscr{E}_0(\theta)$ *and* $(j^*h')^{-1}$ *(generator)* $\subset\pi^k(T(\theta))$.

Proof. Suppose $b_i:\theta\to\theta$, $i=0,1$ and $T(b_0)^*(\beta)=T(b_1)^*(\beta)$ in $\pi^k(T(\theta))$. Let $H:T(\theta)\times I\to S^k$ be a homotopy between them, so that $H(x,i)=T(p_i)$, $i=0,1$, $p_i:E_0(\theta)\to S^{k-1}$ is such that $b_i(x,t)=(x,p_i(x,t))$. Then since $T(\theta)=B\times S^k/B\times *$, $*\in S^k$, we get $B\times S^k\times I\xrightarrow{\eta}T(\theta)\xrightarrow{H}S^k$, and this induces a fibre homotopy between the images of b_0 and b_1 in $\mathscr{E}(\theta+\varepsilon^1)$. Since $k>>\dim B$, it follows from (I.4.12) that $\{b_0\}=\{b_1\}$ in $\mathscr{E}(\theta)$.

If $\beta\in\pi^k(T(\theta))$ is such that $j^*h'(\beta)=$ generator of $H^k(S^k)$, then the composite $B\times S^k\xrightarrow{\eta}T(\theta)\xrightarrow{\beta}S^k$ is of degree 1 on each fibre, so $b(x,t)=(x,\beta\eta(x,t))$ is a fibre homotopy equivalence, so the map is onto $(j^*h')^{-1}$(generator). □

Let ν be the Spivak normal fibre space over a Poincaré pair of dimension m, $\alpha\in\pi_{m+k}(T(\nu)/T(\nu|Y))$ such that $h(\alpha)\cap U=[X]$, $(k>>m)$. If $b:\nu\to\nu$ is a fibre homotopy equivalence of ν with itself, then $\alpha'=T(b)_*(\alpha)$ has the same property, i.e. $h(\alpha')\cap U=[X]$.

I.4.18 Corollary. *The mapping* $\psi : \mathscr{E}_0(v) \to \pi_{m+k}(T(v)/T(v|Y))$ *given by* $\psi(b) = T(b)_*(\alpha)$ *establishes a* $1-1$ *correspondence between* $\mathscr{E}_0(v)$ *and the subset* $K \subset \pi_{m+k}(T(v)/T(v|Y))$, $K = \{\beta | h(\beta) \cap U = [X]\}$.

Proof. By (I.4.16) there is a commutative diagram:

$$\begin{array}{ccccc} \mathscr{E}_0(\varepsilon) & \xleftarrow{\gamma'} & \mathscr{E}_0(\theta) & \xrightarrow{\gamma} & \mathscr{E}_0(v) \\ \downarrow{\scriptstyle \eta'} & & & & \downarrow{\scriptstyle \eta} \\ \{A, A\} & & \xrightarrow{\mathscr{D}} & & \{B, B\} \end{array}$$

where $A = T(\varepsilon)$, ε = a trivial bundle, $(\varepsilon = v + (-v))$, $B = T(v)/T(v|Y)$, η, η' give the reduced maps of Thom complexes, $\mathscr{D}$ is the Spanier-Whitehead duality, $\{\ ,\ \}$ denoting homotopy classes of maps in S-theory. Now we have another commutative diagram

$$\begin{array}{ccc} \{A, A\} & \xrightarrow{\mathscr{D}} & \{B, B\} \\ \downarrow{\scriptstyle e'} & & \downarrow{\scriptstyle e} \\ \{A, S^l\} & \xrightarrow{\mathscr{D}'} & \{S^{m+k}, B\} \end{array}$$

where $\mathscr{D}'$ is an isomorphism of groups, from Spanier-Whitehead duality, and $e'(g) = g^*(\beta)$, $e(f) = f_*(\alpha)$, $\alpha \in \{A, S^l\} = \pi^l(T(\varepsilon))$,

$$\beta \in \{S^{m+k}, B\} = \pi_{m+k}(T(v)/T(v|Y)),$$

as above, where β is chosen so that $\mathscr{D}'(\beta) = \alpha$.

By (I.4.17), the composition $e'\eta'\gamma' : \mathscr{E}_0(\theta) \to \{A, S^l\} = \pi^l(T(\varepsilon))$ is a $1-1$ correspondence onto $h'^{-1}(h'(\beta))$, and since $\mathscr{D}'$ is an isomorphism it follows that $e\eta\gamma : \mathscr{E}_0(\theta) \to \{S^{m+k}, T(v)/T(v|Y)\}$ is a $1-1$ correspondence on $h^{-1}(h(\alpha))$, and since γ is an isomorphism by (I.4.6), the result follows. □

Now we may prove the uniqueness of the Spivak normal fibre space, in the enriched version of [67].

I.4.19 Theorem. *Let* ξ_1 *and* ξ_2 *be* $(k-1)$*-spherical fibre spaces over a Poincaré pair* (X, Y) *of dimension* m, $k >> m$. *Let* $\alpha_i \in \pi_{m+k}(T(\xi_i)/T(\xi_i|Y))$, $i = 1, 2$ *be such that* $h(\alpha_i) \cap U_i = [X]$. *Then there is a fibre homotopy equivalence* $b : \xi_1 \to \xi_2$ *such that* $T(b)_*(\alpha_1) = \alpha_2$, *and such a* b *is unique up to fibre homotopy.*

I.4.20 Lemma. ξ_1 *and* ξ_2 *are fibre homotopy equivalent.*

Proof. By (I.4.15), if v is the Spivak normal fibre space of (X, Y) then if $\xi = \xi_i$, $T(\xi)/T(\xi|Y)$ is S-dual to $T(v + (-\xi))$. Since $\alpha \in \pi_{m+k}(T(\xi)/T(\xi|Y))$ such that $h(\alpha) \cap U$, it follows that $\mathscr{D}'(\alpha) = \beta \in \pi^l(T(v + (-\xi))$ is such that $j^* h'(\beta)$ is a generator of $H^l(S^l)$, and hence the composite

$$E_0(v + (-\xi) + \varepsilon^1) \to T(v + (-\xi)) \to S^l$$

defines a fibre homotopy trivialization of $\nu+(-\xi)+\varepsilon^1$, so that $\xi=\xi_i$ is fibre homotopy equivalent to $\nu+\varepsilon^1$. Hence ξ_1 is fibre homotopy equivalent to ξ_2. □
fibre homotopy to $\nu+\varepsilon^1$. Hence ξ_1 is fibre homotopy equivalent to ξ_2. □

Proof of (I.4.19). By (I.4.20), there is a fibre homotopy equivalence $b_1:\xi_1\to\xi_2$. By (I.4.18), there is a fibre homotopy equivalence $b_2:\xi_2\to\xi_2$ such that $T(b_2)_*(T(b_1)_*(\alpha_1))=\alpha_2$. Hence $T(b_2b_1)_*(\alpha_1)=\alpha_2$.

If $b_1, b_2:\xi_1\to\xi_2$ are two fibre homotopy equivalences such that $T(b_1)_*(\alpha_1)=T(b_2)_*(\alpha_1)=\alpha_2$, then $T(b_2^{-1}b_1)_*(\alpha_1)=\alpha_1$. By (I.4.18), $b_2^{-1}b_1$ is fibre homotopic to the identity, and hence b_1 is fibre homotopic to b_2. □

II. The Main Results of Surgery

In this chapter we shall try to give the main results of the theory of surgery on simply-connected manifolds and give some of the most general and important theorems on the structure of differentiable manifolds which result.

In § 1 we give without proof the main technical results of surgery, the proofs being given in Chapters III, IV and V. They are all stated without reference to surgery as such, but in terms of "normal cobordism" of "normal maps" which are defined in § 1. Surgery is a process used to construct normal cobordisms. In § 2 we discuss some generalities about the relation of normal maps and cobordisms to homotopy groups using transversality. In § 3 we give some of the main theorems on the homotopy type of manifolds and the classification of manifolds. In § 4 we describe a dual approach, which gives the classification theorem a more functorial form.

§ 1. The Main Technical Results

Let (X, Y) be a Poincaré pair of dimension m (see I § 2), where Y may be empty. Let $(M, \partial M)$ be a smooth compact oriented m-manifold with boundary, and let $f:(M, \partial M) \to (X, Y)$ be a map.

A *cobordism* of f, is a pair (W, F) where W^{m+1} is a smooth compact $(m+1)$ manifold, $\partial W = M^m \cup U^m \cup M'^m$, $\partial U = \partial M \cup \partial M'$, $F:(W, U) \to (X, Y)$ and $F|M = f$. If $U = \partial M \times I$ and $F(x, t) = f(x)$ for $x \in \partial M$, $t \in I$, then (W, F) will be called a *cobordism* of f *rel* Y. If we pick a function $\varrho: W \to [0, 1]$ such that $\varrho(M) = 0$, $\varrho(M') = 1$, then

$$G = (F \times \varrho): (W, U) \to (X \times [0, 1], Y \times [0, 1]),$$

and G can be considered

$$G: (W, \partial W) \to (X \times [0, 1], X \times 0 \cup Y \times [0, 1] \cup X \times 1).$$

If (W, F) is a cobordism rel Y, ϱ can be chosen so that $G(x, t) = (f(x), t)$, $x \in \partial M$, $t \in I$.

Let us assume that $k>>m$, and that $(M^m, \partial M)$ is embedded in (D^{m+k}, S^{m+k-1}) with normal bundle ν^k, so that $\nu|\partial M =$ normal bundle of ∂M in S^{m+k-1}.

Let ξ^k be a k-plane bundle over X, $k>>m$. A *normal map* is a map of degree 1, $f:(M,\partial M)\to(X,Y)$ together with a bundle map $b:\nu^k\to\xi^k$ covering f. A *normal cobordism* (W,F,B) of (f,b) is a cobordism (W,F) of f, together with an extension $B:\omega^k\to\xi^k$ of b where ω^k is the normal bundle of W^{m+1} in $D^{m+k}\times I$, where

$$(M,\partial M)\subset(D^{m+k}\times 0, S^{m+k-1}\times 0), \qquad (M',\partial M')\subset(D^{m+k}\times 1, S^{m+k-1}\times 1),$$

and $U^m\subset S^{m+k-1}\times I$.

A normal cobordism *rel* Y, is a cobordism rel Y such that it is a normal cobordism and $B(v,t)=b(v)$ for $v\in\nu|\partial M$, $t\in I$.

Now we can state the surgery problem:

Problem. Given a normal map (f,b), $f:(M,\partial M)\to(X,Y)$, $b:\nu^k\to\xi^k$, when is (f,b) normally cobordant to a homotopy equivalence of pairs?

We may also state:

Restricted Problem. Given a normal map (f,b), $f:(M,\partial M)\to(X,Y)$, $b:\nu\to\xi$, when is (f,b) normally cobordant rel Y to (f',b') where $f':M'\to X$ is a homotopy equivalence?

Of course, if $\partial M=\emptyset$, the restricted problem is the same as the unrestricted one.

II.1.1 Invariant Theorem. *Let (f,b) be a normal map, $f:(M^m,\partial M^m)\to(X,Y)$ etc., such that $f|\partial M$ induces an isomorphism on homology. There is an invariant $\sigma(f,b)$ defined, $\sigma=0$ if m is odd, $\sigma\in\mathbb{Z}$ if $m=4k$ and $\sigma\in\mathbb{Z}_2$ if $m=4k+2$, and such that $\sigma(f,b)=0$ if (f,b) is normally cobordant to a map inducing isomorphism on homology.*

Actually σ will be defined in more generality for normal maps of Poincaré pairs (see Chapter III).

II.1.2 Fundamental Surgery Theorem. *Let (f,b) be a normal map $f:(M,\partial M)\to(X,Y)$, $b:\nu\to\xi$ as usual and suppose*

(1) *$f|\partial M$ induces an isomorphism in homology*

(2) *X is 1-connected*

(3) *$m\geqq 5$.*

If m is odd then (f,b) is normally cobordant rel Y *to a homotopy equivalence $f':M'\to X$. If $m=2k$, (f,b) is normally cobordant* rel Y *to (f',b') such that $f':M'\to X$ is a homotopy equivalence if and only if $\sigma(f,b)=0$.*

This theorem is contained essentially in the work of Kervaire-Milnor [34], Novikov [49], [50], and the author [6].

Kervaire-Milnor proved but did not publish the following (see also [31]):

II.1.3 Plumbing Theorem. *Let* $(X, Y)=(D^m, S^{m-1})$. *If* $m=2k>4$, *then there are normal maps* (g, c), $g:(M, \partial M)\to(D^m, S^{m-1})$, $c: v^k\to\varepsilon^k$ $=$ *trivial bundle, with* $g|\partial M$ *a homotopy equivalence and with* $\sigma(g, c)$ *taking on any desired value.*

It is proved by a technique called "plumbing", and hence its name.

For applications, we will need several properties of the invariant σ.

Let (f, b), $f:(M, \partial M)\to(X, Y)$, etc. be a normal map. Suppose (X, Y) is the sum of two Poincaré pairs (see (I.3.2)), i.e.

$$X=X_1\cup X_2,\ X_0=X_1\cap X_2,\ Y_i=X_i\cap Y, \quad i=0, 1, 2,$$

and $(X_i, \partial X_i)$ where $\partial X_i=X_0\cup Y_i$, are Poincaré pairs $i=1, 2$, with orientation compatible with that of X.

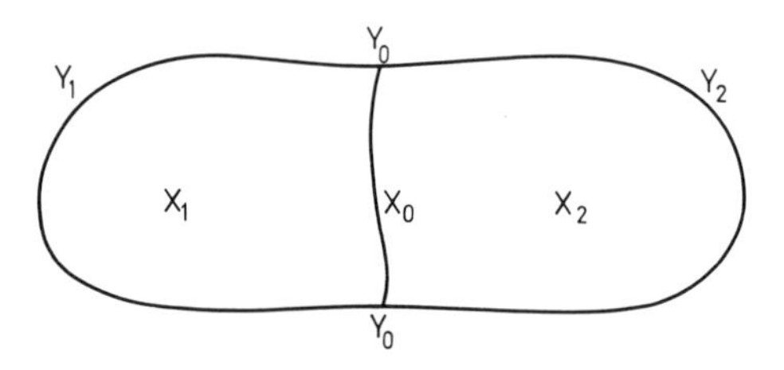

Suppose also that

$$M^m=M_1^m\cup M_2^m, M_0=M_1\cap M_2\subset\partial M_1\cap\partial M_2, \partial M\cap M_i\subset\partial M_i, (M_i, \partial M_i)$$

are compact smooth manifolds with boundary $i=0, 1, 2$. Suppose further that $f(M_i)\subset X_i$, and set $f_i=f|M_i:(M_i, \partial M_i)\to(X_i, \partial X_i)$, $i=1, 2$. Since $v_i=v|M_i$ is the normal bundle of $M_i\subset D^{m+k}$, $i=1, 2$, if $b_i=b|v_i$, then (f_i, b_i) are normal maps, $i=1, 2$. We will say (f, b) is the sum of (f_1, b_1) and (f_2, b_2). If $f|\partial M$ and $f_i|\partial M_i$, $i=1, 2$ induce isomorphisms in homology then σ is defined for each map.

II.1.4 Addition Property. *Suppose* (f, b) *is a normal map which is the sum of two normal maps* (f_1, b_1) *and* (f_2, b_2) *as above, and such that* $f|\partial M, f|\partial M_i$, $i=1$, 2 *and* $f|M_0$ *induce isomorphisms in homology. Then*

$$\sigma(f, b)=\sigma(f_1, b_1)+\sigma(f_2, b_2).$$

II.1.5 Cobordism Property. *Let* (f, b) *be a normal map*

$$f:(M, \partial M)\to(X, Y), \quad b: v\to\xi,$$

and set $f'=f|\partial M:\partial M\to Y$, $b'=b|(v|\partial M): v|\partial M\to\xi|Y$. *If* $m=2k+1$, *then* $\sigma(f', b')=0$.

II.1.6 Index Property. *If* $Y=\emptyset$, $m=4k$, (f, b) *a normal map, then* $8\sigma(f, b)=\text{index } M-\text{index } X$, *and* $\text{index } M=L_k(p_1(\xi^{-1}), \ldots) [X]$, *where* L_k *is the Hirzebruch polynomial, and* $\text{index } X=$ *signature of the quadratic form on* $H^{2k}(X; \mathbb{Q})$ *given by* $\langle x\cup y, [X]\rangle$, *where* $[X]$ *is the orientation class of* $H_{4k}(X)$.

II.1.7 Product Formulas. *Let* (f_1, b_1), (f_2, b_2) *be normal maps,* $f_i:(M_i, \partial M_i)\to(X_i, \partial X_i)$. *Suppose* $\sigma(f_1\times f_2, b_1\times b_2)$, $\sigma(f_1, b_1)=\sigma_1$ *and* $\sigma(f_2, b_2)=\sigma_2$ *are all defined* (i.e. $f_1\times f_2|\partial(M_1\times M_2)$, $f_i|\partial M_i$, $i=1, 2$ *are all homology isomorphisms with appropriate coefficients*). *Then*

(i) $\sigma(f_1\times f_2, b_1\times b_2)=I(X_1)\sigma_2+I(X_2)\sigma_1+8\sigma_1\sigma_2$ *when dimension* $M_1\times M_2=4k$, *where* $I(X_i)$ *is the index of* X_i.

(ii) $\sigma(f_1\times f_2, b_1\times b_2)=\chi(X_1)\sigma_2+\chi(X_2)\sigma_1$, *when dimension* $M_1\times M_2=4k+2$, *where* $\chi(X_i)=$ *Euler characteristic of* X_i.

Note that $I(X)=0$ by definition if $\dim X\not\equiv 0 \pmod 4$.

The three theorems and the four properties of the obstruction σ are the main technical results in the theory of surgery on simply-connected manifolds. In the next sections we will show how to deduce some of the main theorems of the subject from these results, and we will prove the technical results in later chapters.

§ 2. Transversality and Normal Cobordism

In this section we recall the transversality results due to Thom which we shall need, and derive from them the relation between normal cobordism classes and homotopy classes of maps.

Let Z be a space and suppose there is a vector bundle ξ^k with base space X embedded as an open set in Z, where X is the homotopy type of a finite complex. Let M^n be a differential manifold and $f: M\to Z$ a continuous map.

We shall say that f is *transversal* to $X\subset Z$ if $f^{-1}X=N^{n-k}$ a smooth submanifold of M with normal bundle ν^k, and f restricted to a tubular neighborhood of N in M is a linear bundle map of ν into ξ.

This definition is usually given as a theorem which follows from the usual notion of transversality, but this is exactly what we need so we use it as the definition. The proof of the following theorem, which is due to Thom, may be extracted from many standard treatments (see [64], [1]). The fact that X is not a smooth manifold makes no difference; for example one could replace X by an open subset of euclidean space, by taking a regular neighborhood, and get a Z of the same homotopy type, or on the other hand one notes that none of the arguments of the transversality theorems use differentiability in the base, but only in the fibre.

II.2.1 Thom Transversality Theorem. *Let A be closed in M and suppose that f restricted to an open neighborhood of A is already transversal to X. Then there is a homotopy of f rel A to f' such that f' is transversal to X.*

Actually this homotopy can be taken to be very small in some metric, but such refinements will not concern us.

Suppose ξ^k is a linear k-plane bundle over X. We recall the definition of the Thom complex $T(\xi)$ of the bundle $\xi: T(\xi)=E(\xi)/E_0(\xi)$, where $E(\xi)$ is the closed disk bundle associated with ξ, i.e. with fibre D^k, and $E_0(\xi)$ is the associated sphere bundle, i.e. with fibre S^{k-1}, so that $E_0(\xi)\subset E(\xi)$.

Recall that $\xi+\varepsilon^1$, where ε^1 is the trivial line bundle, has as total space the total space of ξ times R^1. Thus $E(\xi+\varepsilon^1)=E^+(\xi+\varepsilon^1)\cup E^-(\xi+\varepsilon^1)$ and $E_0(\xi+\varepsilon^1)=E_0^+(\xi+\varepsilon^1)\cup E_0^-(\xi+\varepsilon^1)$ where $E^+(\xi+\varepsilon^1)$ is the subset where the coordinate t in R^1 is $\geqq 0$, $E^-(\xi+\varepsilon^1)$ where $t\leqq 0$, etc. Then $E(\xi)\subset E(\xi+\varepsilon^1)$ as the subset where $t=0$, and $E(\xi)=E^+(\xi+\varepsilon^1)\cap E^-(\xi+\varepsilon^1)$ and $E_0(\xi)=E_0^+(\xi+\varepsilon^1)\cap E_0^-(\xi+\varepsilon^1)$. Also the projection of total space on the first factor (i.e., forgetting the coordinate t in R^1) induces a map $p: E(\xi+\varepsilon^1)\to E(\xi)$ such $p'=p\,|\,E_0^+(\xi+\varepsilon^1): E_0^+(\xi+\varepsilon^1)\to E(\xi)$ is a homeomorphism, and $p'\,|\,E_0^+(\xi+\varepsilon^1)\cap E_0^-(\xi+\varepsilon^1)$ is a homeomorphism (the identity) onto $E_0(\xi)$. For in each fibre we note that p looks like the projection of a disk onto a disk one dimension lower:

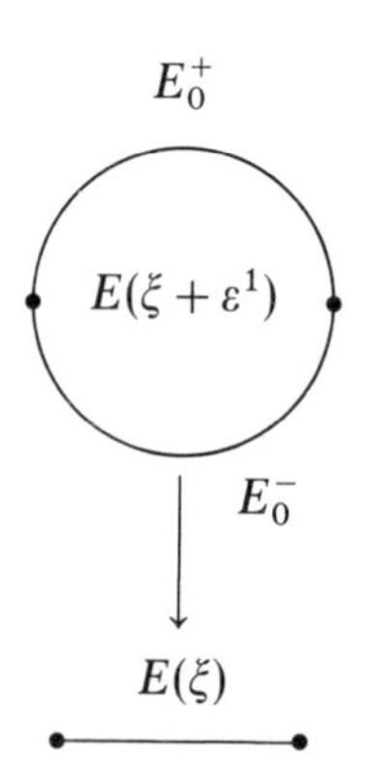

Hence it follows that

$$
\begin{aligned}
T(\xi)=E(\xi)/E_0(\xi)&=E_0^+(\xi+\varepsilon^1)/E_0^+(\xi+\varepsilon^1)\cap E_0^-(\xi+\varepsilon^1)\\
&=E_0(\xi+\varepsilon^1)/E_0^-(\xi+\varepsilon^1)
\end{aligned}
$$

(where $=$ means homeomorphic).

Now let $\alpha: X\to E_0(\xi+\varepsilon^1)$ be the canonical cross-section where $\alpha(x)=(0,-1)$ in the fibre over $x\in X$, i.e. the unique point with t coordinate $=-1$. Let $\varrho:[0,1]\to[-1,1]$ by $\varrho(t)=2t-1$, so that ϱ is a homeomorphism. Then define $\varrho' : E_0^+(\xi+\varepsilon^1)\to E_0(\xi+\varepsilon^1)$ by

$$\varrho'(b,t)=\left(\frac{\sqrt{1-\varrho(t)^2}}{|b|}\,b,\varrho(t)\right),$$

for $b\in E(\xi)$ such that $|b|^2+|t|^2=1$, so that $(b,t)\in E_0(\xi+\varepsilon^1)$ and $|b|\neq 0$. If $|b|=0$, then $t=1$, and we define $\varrho'(b,1)=(b,1)$. It is then clear that ϱ' induces a homeomorphism between $E_0^+(\xi+\varepsilon^1)/E_0^+(\xi+\varepsilon^1)\cap E_0^-(\xi+\varepsilon^1)$ and $E_0(\xi+\varepsilon^1)/\alpha(X)$, so that

II.2.2 Proposition. *$T(\xi)$ is homeomorphic to $E_0(\xi+\varepsilon^1)/\alpha(X)$.*

II.2.3 Thom Isomorphism Theorem. *Let ξ^k be a linear k-plane bundle over a connected space X, and let $j:(D^k,S^{k-1})\to(E(\xi),E_0(\xi))$ be the map induced by the inclusion of a fibre D^k into $E(\xi)$, $\pi:E(\xi)\to X$. If $U\in H^k(E(\xi),E_0(\xi))$ is such that j^*U generates $H^k(D^k,S^{k-1})$ then $\Phi:H^q(X)\to H^{q+k}(E(\xi),E_0(\xi))$ is an isomorphism, where $\Phi(x)=U\cup\pi^*(x)$. Further there is always such a U with $\mathbb{Z}_2$ coefficients, (i.e. with $H^*=H^*(\ ;\mathbb{Z}_2)$ in the statement) and the existence of such a $U\in H^*(E(\xi),E_0(\xi);\mathbb{Z})$ is equivalent to the orientability of ξ.*

The element U will be called the *Thom class* of ξ.
Here the $\cup$ product is the relative one

$$H^*(E(\xi),E_0(\xi))\otimes H^*(E(\xi))\to H^*(E(\xi),E_0(\xi)).$$

This theorem has many modern proofs, for example using a spectral sequence (see (I.4.3)) or using a Mayer-Viettoris theorem (see [44], and also [32]). One can use (II.2.2) and standard methods of studying $H^*(E_0(\xi+\varepsilon^1))$, for example. The theorem holds in more generality for spherical fibre spaces, and also has a converse in this context, essentially due to Spivak (see (I.4.3)).

Since the collapsing map $\varrho:(E(\xi),E_0(\xi))\to(T(\xi),*)$ induces an isomorphism in cohomology, we get as an easy corollary:

II.2.4 Corollary. *If ξ is orientable, (or otherwise with $\mathbb{Z}_2$ coefficients) the map $\Phi:H^q(X)\to\bar{H}^{q+k}(T(\xi))$ given by $\Phi(x)=\varrho^{*-1}(x\cup U)$ is an isomorphism, for all q.*

($\bar{H}^*$ denotes reduced cohomology.)

Let $Y\subset X$ be a closed subset and let $\xi'=\xi\,|\,Y$. Then a relative cup product is naturally defined:

$$H^q(E(\xi),E_0(\xi))\otimes H^l(E(\xi),E(\xi'))\to H^{q+l}(E(\xi),E_0(\xi)\cup E(\xi')).$$

The projection $\pi:(E(\xi),E(\xi'))\to(X,Y)$ induces an isomorphism in cohomology, as does the natural collapse of $E_0(\xi)$ which defines

$$\eta:(E(\xi),E_0(\xi)\cup E(\xi'))\to(T(\xi),T(\xi')).$$

Hence we get a cup product

$$\bar{H}^q(T(\xi)) \otimes H^l(X, Y) \to H^{q+l}(T(\xi), T(\xi')).$$

II 2.5 Corollary. *If ξ is orientable (or otherwise with $\mathbb{Z}_2$ coefficients) the map $\Phi: H^q(X, Y) \to H^{q+k}(T(\xi), T(\xi'))$ given by $\Phi(x) = U \cup x$, is an isomorphism, for all q.*

The proof is a simple application of the Five Lemma, using the commutative diagram:

$$\begin{array}{ccccccccc}
\cdots \to & H^q(X) & \to & H^q(Y) & \to & H^{q+1}(X, Y) & \to & H^{q+1}(X) & \to \cdots \\
& \downarrow{\scriptstyle \cup U} & & \downarrow{\scriptstyle \cup U'} & & \downarrow{\scriptstyle \cup U} & & \downarrow & \\
\cdots \to & H^{q+k}(T(\xi)) & \xrightarrow{g^*} & H^{q+k}(T(\xi')) & \to & H^{q+k+1}(T(\xi), T(\xi')) & \to & H^{q+k+1}(T(\xi)) & \to \cdots
\end{array}$$

where $U' = g^*(U)$. Here $g: T(\xi') \to T(\xi)$ is the inclusion, and it follows that $U' = g^*(U)$ has the property that restricted to the fibre, $j'^* g^* U = j^* U$ generates $\bar{H}^k(S^k)$, where $j': S^k \to T(\xi')$, $j: S^k \to T(\xi)$ come from the inclusions of the fibres, so $gj' = j$. Hence, by (II.2.4), the maps on the spaces are isomorphisms, so by the Five Lemma the map of pairs is an isomorphism. (Here note that the restriction of an orientable bundle is orientable.)

Using the relation between $\cup$ and $\cap$ product developed in Chapter I, we may easily derive homology versions of the Thom isomorphisms:

II.2.6 Theorem. *With the hypotheses of* (II.2.4), *the map $\cap U: H_{q+k}(T(\xi)) \to H_q(X)$ is an isomorphism for all q.*

II.2.7 Theorem. *With the hypotheses of* (II.2.5), *the map $\cap U: H_{q+k}(T(\xi), T(\xi')) \to H_q(X, Y)$ is an isomorphism for all q.*

Proof. Considering the orientable case, $\cap U$ is induced by a chain map (see Chapter I), so it follows that $\cap U$ is an isomorphism if and only if $\cap U: H_{q+k}(T(\xi); \mathbb{Z}_p) \to H_q(X; \mathbb{Z}_p)$ is an isomorphism for all q, all primes p. (In the non-orientable case, we only consider $\mathbb{Z}_2$ anyway.) Let $y \in H_{q+k}(T(\xi); \mathbb{Z}_p)$. If $x \in H^q(X; \mathbb{Z}_p)$, then by (I.1.1) using the evaluation of cohomology on homology, $x(y \cap U) = (x \cup U)(y)$. Since

$$\cup U: H^q(X; \mathbb{Z}_p) \to H^{q+k}(T(\xi); \mathbb{Z}_p)$$

is an isomorphism, it follows that $x(y \cap U) = 0$ for all x if and only if $y = 0$. Hence $\cap U$ is a monomorphism. Now

$$H^{q+k}(T(\xi); \mathbb{Z}_p) = \operatorname{Hom}(H_{q+k}(T(\xi); \mathbb{Z}_p), \mathbb{Z}_p)$$

and $H^q(X; \mathbb{Z}_p) = \operatorname{Hom}(H_q(X; \mathbb{Z}_p), \mathbb{Z}_p)$ and $H^q(X; \mathbb{Z}_p) \cong H^{q+k}(T(\xi); \mathbb{Z}_p)$ by the Thom isomorphism $\cup U$. Hence the homology groups $H_q(X; \mathbb{Z}_p)$

and $H_{q+k}(T(\xi);\mathbb{Z}_p)$ have the same rank over $\mathbb{Z}_p$, and hence the monomorphism $\cap U$ is an isomorphism. This proves (II.2.6). A similar argument proves (II.2.7). □

Now let $f:(X,Y)\to(X',Y')$ be a map, ξ^k and $\xi^{k'}$ linear k-plane bundles over X and X' respectively, and $b:\xi\to\xi'$ a linear bundle map covering f. Let $T(b):T(\xi)\to T(\xi')$ be the induced map of Thom complexes. Assume either ξ' (and hence ξ) orientable or use $\mathbb{Z}_2$ coefficients below.

II.2.8 Lemma. *$T(b)^*(U')=U$ where $U'\in H^k(T(\xi'))$ is such that j'^*U' generates $H^k(S^k)$ (as above), similarly for U.*

Proof. Since b is linear $bj=j'$, where j, j' are inclusions of fibres, and the result follows. □

II.2.9 Theorem. (Naturality of the Thom Isomorphism). *With the above hypotheses, if Φ and Φ' are the Thom isomorphisms in ξ and ξ' respectively, then $T(b)^*\Phi'=\Phi f^*$, and $T(b)_*(x)\cap U'=f_*(x\cap U)$, $x\in H_{q+k}(T(\xi),T(\xi|Y))$.*

Proof.

$$T(b)^*\Phi'(y)=T(b)^*(U'\cup\pi'^*(y))=T(b)^*(U')\cup b^*\pi'^*(y)=U\cup\pi^*f^*(y)=\Phi f^*(y).$$

Similarly,

$$f_*(x\cap U)=f_*(x\cap T(b)^*U')=T(b)_*(x)\cap U'.$$

Here we may think of the map of pairs $b:(E(\xi),E_0(\xi))\to(E(\xi'),E_0(\xi'))$ instead of $T(b)$ in order to find the necessary identities between $\cup$ and $\cap$ in Chapter I. □

II.2.10 Corollary. *With notation as above, suppose (X,Y) and (X',Y') are Poincaré pairs of dimension n. Then the degree of f is equal to the degree of $T(b)$, in particular, $f_*:H_n(X,Y)\to H_n(X',Y')$ is an isomorphism if and only if $T(b)_*:H_{n+k}(T(\xi),T(\xi|Y))\to H_{n+k}(T(\xi'),T(\xi'|Y'))$ is an isomorphism.*

Proof. If $[X]\in H_n(X,Y)$, $[X']\in H_n(X',Y')$, $v\in H_{n+k}(T(\xi),T(\xi|Y))$, $v'\in H_{n+k}(T(\xi'),T(\xi'|Y'))$ are generators such that $v\cap U=[X]$, $v'\cap U'=[X']$, then from (II.2.9), $f_*[X]=(T(b)_*(v))\cap U'$ and the result follows. □

Let X be a space, ξ^k a k-plane bundle over X, Y closed subset of X, and suppose the total space $E(\xi)$ of the associated unit disk (D^k) bundle is embedded as a closed subset of a space Z, such that interior of $E(\xi)$ is open in Z, and such that $E(\xi|Y)\subset Z'$, a closed subset of Z. Hence $Z=E(\xi)\cup A$, $Z'=E(\xi|Y)\cup A'$, where

$$E(\xi)\cap A=E_0(\xi),\ E(\xi|Y)\cap A'=E_0(\xi|Y),\ A'=A\cap Z'.$$

II.2.11 Definition. We define the natural collapsing map $\eta:(Z,Z')\to(T(\xi),T(\xi|Y))$ extending the identity map of $E(\xi)$, such that $\eta(A)=*$.

If X is a smooth manifold with boundary Y embedded with normal bundle ξ^k in a smooth manifold Z with boundary Z' then we get such a situation. In that case we get:

II.2.12 Lemma. *Suppose* $(M,\partial M)$ *is an oriented smooth manifold of dimension* n *embedded with normal bundle* v^k *in an oriented smooth* $(n+k)$*-manifold* W*, with* $\partial M\subset\partial W$*. Then the natural collapse*

$$\eta:(W,\partial W)\to(T(v),T(v|\partial M))$$

has degree 1, i.e. $\eta_*:H_{n+k}(W,\partial W)\to H_{n+k}(T(v),T(v|\partial M))$ *is an isomorphism,* $\eta_*[W]\cap U=[M]$ *for the appropriate choice of* $U\in H^k(T(v))$.

This is geometrically clear, and can be shown purely algebraically by looking at an appropriate diagram.

Let (X,Y) be a Poincaré pair of dimension n, ξ^k an oriented linear k-plane bundle over X, $k>n$. Let (f,b) be a normal map so that $f:(M,\partial M)\to(X,Y)$ is a map of degree 1, M a smooth oriented n-manifold with boundary, v^k its normal k-plane bundle in (D^{n+k},S^{n+k-1}), $b:v\to\xi$ a linear bundle map covering f. Then b induces a map of Thom complexes $T(b):(T(v),T(v|\partial M))\to(T(\xi),T(\xi|Y))$. Let

$$\eta:(D^{n+k},S^{n+k-1})\to(T(v),T(v|\partial M))$$

be the natural collapse, and consider the composite

$$T(b)\eta:(D^{n+k},S^{n+k-1})\to(T(\xi),T(\xi|Y)).$$

The homotopy class of $T(b)\eta$ in $\pi_{n+k}(T(\xi),T(\xi|Y))$ will be called the *Thom invariant* of the normal map (f,b).

II.2.13 Theorem. *The Thom invariant of* (f,b) *depends only on the normal cobordism class of* (f,b)*, and defines a* $1-1$ *correspondence between normal cobordism classes of normal maps, and elements* $\alpha\in\pi_{n+k}(T(\xi),T(\xi|Y))$ *such that* $h(\alpha)\cap U=[X]\in H_n(X,Y)$*, where* $h:\pi_{n+k}\to H_{n+k}$ *is Hurewicz homomorphism, and* $[X]$ *is the orientation class,* $U\in H^k(T(\xi))$ *is the Thom class for* ξ.

Proof. Suppose (f,b) is normally cobordant to (f',b') so that there is a manifold W^{n+1} with $\partial W=M\cup V\cup M'$, $\partial V=\partial M\cup\partial M'$, a map $F:(W,V)\to(X,Y)$ with $F|M=f$, $F|M'=f'$, and $B:\omega\to\xi$ a linear bundle map, where ω is the normal k-plane bundle of

$$(W,V)\subset(D^{n+k}\times I,S^{n+k-1}\times I),\quad W\cap D^{n+k}\times 0=M,\quad W\cap D^{n+k}\times 1=M',$$

and the restrictions of B to the two ends give b and b' respectively. Then if $\eta:(D^{n+k}, S^{n+k-1})\to(T(\nu), T(\nu|\partial M))$, $\eta':(D^{n+k}, S^{n+k-1})\to(T(\nu'), T(\nu'|\partial M'))$ and $\zeta:(D^{n+k}\times I, S^{n+k-1}\times I)\to(T(\omega), T(\omega|V))$ it follows easily that $T(B)\zeta|D^{n+k}\times 0=T(b)\eta$ and $T(B)\zeta|D^{n+k}\times 1=T(b')\eta'$, so that the Thom invariants are homotopic.

Now let $\alpha\in\pi_{n+k}(T(\xi), T(\xi|Y))$ such that $h(\alpha)\cap U=[X]$. Let $f:(D^{n+k}, S^{n+k-1})\to(T(\xi), T(\xi|Y))$ represent α and by the Thom Transversality Theorem (II.2.1) we may assume f is transverse to X and Y in $T(\xi)$ and $T(\xi|Y)$, so that $f^{-1}(X, Y)=(M, \partial M)$ a smooth n-manifold with boundary and f restricted to a tubular neighborhood of X or Y is a linear bundle map b of the normal bundle ν of $(M, \partial M)\subset(D^{n+k}, S^{n+k-1})$ into ξ. Now if we take $g=f|(M, \partial M)$, we claim that $g_*([M])=[X]$, where $[M]=\eta_*(\iota)\cap U_\nu$, where $\eta:(D^{n+k}, S^{n+k-1})\to(T(\nu), T(\nu|\partial M))$ is the collapse, $\iota\in H_{n+k}(D^{n+k}, S^{n+k-1})$ is the generator (for a fixed orientation of D^{n+k}) and $U_\nu\in H^k(T(\nu))$ is the Thom class of the bundle ν, $U_\nu=T(b)^*(U)$. For we have that

$$\begin{aligned} g_*([M]) &= g_*(\eta_*(\iota)\cap U_\nu)=g_*(\eta_*(\iota)\cap T(b)^*(U)) \\ &= T(b)_*\eta_*(\iota)\cap U=f_*(\iota)\cap U \\ &= h(\alpha)\cap U=[X] \end{aligned}$$

using (II.2.9). Hence (g, b) is a normal map and the Thom invariant map is onto $\pi_{n+k}(T(\xi), T(\xi|Y))$.

Now suppose the Thom invariants of two normal maps (f, b) and (f', b') are the same, so that there is a homotopy

$$H:(D^{n+k}\times I, S^{n+k-1}\times I)\to(T(\xi), T(\xi|Y))$$

between $T(b)\eta$ and $T(b')\eta'$. Using the Transversality Theorem (II.2.1) again, we may change H, leaving it fixed on $D^{n+k}\times 0$ and $D^{n+k}\times 1$ so that it is transversal to X and Y, and then it follows that the inverse image of (X, Y) under this new map is a normal cobordism between (f, b) and (f', b'). □

§ 3. Homotopy Types of Smooth Manifolds and Classification

Let us denote by $h:\pi_i\to H_i$ the Hurewicz homomorphism, and if ξ^k is a linear oriented k-plane bundle over X, let $U\in H^k(T(\xi))$ be its Thom class, so that $\cap U: H_{q+k}(T(\xi))\to H_q(X)$ and

$$\cap U: H_{q+k}(T(\xi), T(\xi|Y))\to H_q(X, Y)$$

are isomorphisms ($Y\subset X$), (see (II.2.6), (II.2.7)).

The following theorem is due independently to Novikov [50] and the author [6].

II.3.1 Theorem. *Let X be a 1-connected Poincaré complex of dimension $m \geqq 5$, ξ an oriented linear k-plane bundle over X, $k > m+1$, $\alpha \in \pi_{m+k}(T(\xi))$ such that $h(\alpha) \cap U = [X]$, $U \in H^k(T(\xi))$ the Thom class, $[X] \in H_m(X)$ the orientation class. If*

(i) *m is odd, or*

(ii) *$m = 4k$ and $\operatorname{Index} X = (L_k(p_1(\xi^{-1}), \ldots, p_k(\xi^{-1})))\,[X]$,*

then there is a homotopy equivalence $f: M^m \to X$, M^m a smooth m-manifold, such that $\nu = f^*(\xi)$, ν = normal bundle of $M^m \subset S^{m+k}$, and f can be found in the normal cobordism class represented by α.

Proof. By (II.2.13), there is a normal map (f, b), $f: M \to X$ such that α = Thom invariant of (f, b). By the Fundamental Theorem (II.1.2), (f, b) is normally cobordant to a homotopy equivalence if m is odd, and if $m = 2q$ then (f, b) is normally cobordant to a homotopy equivalence if and only if $\sigma(f, b) = 0$. If $m = 4k$, by the Index Property (II.1.6), $\sigma(f, b) = (L_k(p_1(\xi^{-1}), \ldots,))\,[X]$-index X which $= 0$ when (ii) holds. □

If $m = 4k+2$, it may be difficult to evaluate $\sigma(f, b)$.

II.3.2 Remark. If $m = 6$, 14, 30 or 62, then with the above hypotheses there is a homotopy equivalence $f: M^m \to X$, with $f^*(\xi) = \nu$ as above, but f may not be representable by a normal map with Thom invariant α.

Define the connected sum of normal maps of manifolds: Let (f_1, b_1) and (f_2, b_2) be normal maps, $f_i: M_i^m \to X_i$. Let $M_i^0 = M_i - \operatorname{int} D_i^m$, D_i^m an m-cell in M_i, and let $X_i^0 \subset X$ be a subcomplex such that $X_i = X_i^0 \cup D_i^{m\prime}$, $D_i' \cap \partial X_i = \emptyset$ and $H_m(X_i^0, \partial X_i) = 0$. It is an easy exercise to find a representation of X_i of this type. We may assume using the homotopy extension theorem that $f_i^{-1}(D_i') = D_i$, $f_i \mid D_i^m: D_i^m \to D_i^{m\prime} \subset X_i$, $i = 1, 2$ and if $h: D_1 \to D_2$, $h': D_1' \to D_2'$ are orientation reversing diffeomorphisms, we can arrange that $h'(f_1 \mid D_1) = (f_2 \mid D_2) h$.

Let $M_i^0 = M_i - \operatorname{int} D_i$, $X_i^0 = X_i - \operatorname{int} D_i'$, and define $M_1 \# M_2 = M_1^0 \cup M_2^0$ with ∂D_1 identified to ∂D_2 by $h \mid \partial D_1$, $X_1 \# X_2 = X_1^0 \cup X_2^0$ with $\partial D_1'$ identified to $\partial D_2'$ by $h' \mid \partial D_1'$, and make $M_1 \# M_2$ differentiable. Then the restrictions of f_1 and f_2 to M_1^0 and M_2^0 are compatible with the identifications and define a map $f_1 \# f_2: M_1 \# M_2 \to X_1 \# X_2$. It follows from (I.3.2) that $(X_1 \# X_2, \partial X_1 \cup \partial X_2)$ is a Poincaré pair, and $f_1 \# f_2$ is a map of degree 1. By choosing a bundle equivalence of $\xi_1 \mid D_1'$ with $\xi_2 \mid D_2'$ covering h' we may define $\xi_1 \# \xi_2$, a k-plane bundle over $X_1 \# X_2$, and we may arrange, using the bundle covering homotopy theorem, that b_1 and b_2 are compatible to give a bundle map $b_1 \# b_2: \nu_\# \to \xi_1 \# \xi_2$, where $\nu_\#$ is the normal bundle of $M_1 \# M_2$ in D^{m+k}, $\nu_\# \mid M_i^0 = \nu_i \mid M_i^0$ and $b_1 \# b_2 \mid (\nu_i \mid M_i^0) = b_i \mid (\nu_i \mid M_i^0)$. Then $(f_1 \# f_2,\ b_1 \# b_2)$ is the connected sum of (f_1, b_1) and (f_2, b_2), and it follows from results of [17] and [51], that it is independent of the choices involved.

In case ∂M_i and Y_i are non-empty, (f_i, b_i) normal maps, $f_i:(M_i, \partial M_i)\to(X_i, Y_i)$, we may define the connected sum along components of the boundary as follows: Consider the Euclidean half disk H^m, i.e. $H^m=\{x\in R^m, \|x\|\leqq 1, x_m\geqq 0\}$, so that $D^{m-1}\subset\partial H$. Find differentiable embeddings $(H_i^m, D_i^{m-1})\subset(M_i^m, \partial M_i)$, and define

$$(M_i^0, \partial M_i^0)=(\text{closure } M_i-H_i, \partial M_i-\operatorname{int} D_i).$$

Let $(X_i, Y_i)=(X_i^0, Y_i^0)\cup(H_i', D_i^{m-1\prime})$, such that $H_{m-1}(Y_i^0)=0$, a representation which can be made (if necessary changing (X_i, Y_i) by a homotopy equivalence). Let $D^0=\partial H-\operatorname{int} D^{m-1}$, and let

$$M_1 \amalg M_2 = M_1^0\cup M_2^0$$

with $D_1^0\subset\partial M_1^0$ identified with $D_2^0\subset\partial M_2^0$, by an orientation reversing diffeomorphism. Then $M_1\amalg M_2$ may be made differentiable and $\partial(M_1\amalg M_2)=\partial M_1\#\partial M_2$. One may proceed similarly to the above discussion of the closed case to show that there is defined

$$(f_1\amalg f_2, b_1\amalg b_2),\ f_1\amalg f_2:(M_1\amalg M_2, \partial(M_1\amalg M_2))\to(X_1\amalg X_2, Y_1\# Y_2), \text{etc.}$$

Then this is a sum of normal maps which is exactly the situation in (II.1.4), where the intersection of the two parts (M_0 and X_0 in notation of (II.1.4)) are $(m-1)$-cells.

II.3.3 Lemma. $(X\amalg D^m, Y\# S^{m-1})=(X, Y)$.

The proof is obvious. □

II.3.4 Proposition. *Let (f, b), (g, c) be normal maps,*

$$f:(M^m, \partial M)\to(X, Y),\quad g:(N, \partial N)\to(D^m, S^{m-1}),\quad \text{etc.}$$

Then $(f\amalg g, b\amalg c)$ is normally cobordant to (f, b).

Proof. By (II.3.3), we may assume $f\amalg g:(M\amalg N, \partial(M\amalg N))\to(X, Y)$, and since D^m is contractible, we may assume $(f\amalg g)(N^0)\subset Y$. Take $W=(M\amalg N)\times I$ and define $U\subset\partial W$ as follows

$$U=(N^0\times 0)\cup(\partial(M\amalg N)\times I).$$

Then

$$\partial W=M\cup U\cup(M\amalg N),\ (f\amalg g)\,p_1(U)\subset Y,\ (p_1:(M\amalg N)\times I\to M\amalg N),$$

and it is not hard to see that $b\amalg c$ may be arranged to make W a normal cobordism. □

II.3.5 Proposition. *If (f, b) is a normal map, $f:(M^m, \partial M)\to(X, Y)$, and (h, d) is a normal map $h:(V^{m+1}, \partial V)\to(D^{m+1}, S^m)$, then $(f\#(\partial h), b\#(\partial d))$ is normally cobordant* rel Y *to (f, b) (where $\partial h=h|\partial V$, $\partial d=d|\partial V$).*

Proof. Take $(F, B)=(f\times 1, b\times 1)$.

$$f\times 1:(M\times I, \partial(M\times I))\to(X\times I, X\times 0\cup Y\times I\cup X\times 1).$$

Then (F, B) is a normal map, and if we take $(F\amalg h, B\amalg d)$ along an m-cell in $M\times 1$, the result is a normal cobordism rel Y between (f, b) on $M\times 0$ and $(f\#(\partial h), b\#(\partial d))$ on $(M\times 1)\#(\partial V)$. □

II.3.6 Theorem. *Let (X, Y) be an m-dimensional Poincaré pair with X 1-connected, $Y\neq\emptyset$, $m\geqq 5$, and let (f, b), $f:(M, \partial M)\to(X, Y)$ be a normal map such that $(f|\partial M)_*: H_*(\partial M)\to H_*(Y)$ is an isomorphism. Then there is a normal map (g, c), $g:(U^m, \partial U)\to(D^m, S^{m-1})$, with $(g|\partial U)$ a homotopy equivalence, such that $(f\amalg g, b\amalg c)$ is normally cobordant rel Y to a homotopy equivalence. In particular, (f, b) is normally cobordant to a homotopy equivalence.*

Proof. Let (g, c) be such that $\sigma(g, c)=-\sigma(f, b)$, which exists by (II.1.3), The Plumbing Theorem. By (II.1.4), the Addition Property, $\sigma(f\amalg g, b\amalg c)=\sigma(f, b)+\sigma(g, c)=0$, so by the Fundamental Theorem (II.1.2), $(f\amalg g, b\amalg c)$ is normally cobordant rel Y to (f', b'), where $f': M'\to X$ is a homotopy equivalence. By (II.3.4), (f, b) is normally cobordant to (f', b'). □

Recall that a cobordism W^{m+1} between M^m and M'^m, $\partial W=M\cup U\cup M'$, is called an *h-cobordism* if all the inclusions $M\subset W$, $M'\subset W$, $\partial M\subset U$, and $\partial M'\subset U$ are homotopy equivalences. We recall that Smale [54] has proved that if $m\geqq 5$, U is diffeomorphic to $\partial M\times I$ and W is 1-connected, then the diffeomorphism of U with $\partial M\times I$ and the diffeomorphism $M\to M\times 0$ extend to a diffeomorphism of $W\to M\times I$. In particular M is diffeomorphic to M'.

From this we can deduce the classification theorem of Novikov [49].

II.3.7 Theorem. *Let (f_i, b_i), $i=0, 1$ be normal maps $f_i: M_i^m\to X$, X 1-connected Poincaré complex of dimension $m\geqq 4$, and suppose f_0, f_1 are homotopy equivalences. If f_0 is normally cobordant to f_1, then there is a normal map (g, c), $g:(U^{m+1}, \partial U)\to(D^{m+1}, S^m)$, $(g|\partial U)$ a homotopy equivalence, such that (f_0, b_0) is h-cobordant to $(f_1\amalg g|\partial U, b_1\amalg c|\partial U)$. In particular M_0 is h-cobordant to M_1 if m is even, and to $M_1\#(\partial U)$ if m is odd.*

Proof. Let (F, B), $F:(W, M_0\cup M_1)\to(X\times I, X\times 0\cup X\times 1)$ be the normal cobordism between (f_0, b_0) and (f_1, b_1). Then (F, B) is a normal map, and $F|\partial W=F|M_0\cup M_1$ is a homotopy equivalence by hypothesis, so (II.3.6) applies. Adding (g, c) along M_2, (II.3.6) implies that $(F, B)\amalg(g, c)$ is normally cobordant rel $X\times 0\cup X\times 1$ to (F', B'),

$$F':(W', M_0'\cup M_1')\to(X\times I, X\times 0\cup X\times 1),$$

and $F' : W' \to X \times I$ is a homotopy equivalence, where

$$M'_0 = M_0, M'_1 = M_1 \# \partial U, F' | M'_i = f'_i, f'_0 = f_0, f'_1 = f_1 \# \partial g .$$

Then

$$\begin{array}{ccc} M'_i & \xrightarrow{j_i} & W' \\ {\scriptstyle f'_i}\downarrow & & \downarrow{\scriptstyle F'} \\ X \times i & \longrightarrow & X \times I \end{array}$$

is commutative, f'_i, F' and $X \times i \subset X \times I$ are homotopy equivalences so $j_i : M'_i \subset W'$, $i = 0, 1$ are homotopy equivalences, so W' is an h-cobordism. □

II.3.8 Corollary. *Let M, M' be closed smooth 1-connected manifolds of dimension $\geqq 5$. A homotopy equivalence $f : M \to M'$ is homotopic to a diffeomorphism $f' : M \# \Sigma \to M'$ (where $M = M \# \Sigma$ as a topological space), for some homotopy sphere $\Sigma = \partial U$, U parallelizable, if and only if there is a linear bundle map $b : \nu \to \nu'$ covering f such that $T(b)_* (\alpha) = \alpha'$, α, α' natural collapsing maps $\alpha \in \pi_{m+k}(T(\nu))$, $\alpha' \in \pi_{m+k}(T(\nu'))$.*

Proof. If f is homotopic to such a diffeomorphism f' then $df' : \tau_{M \# \Sigma} \to \tau_{M'}$ induces a map of normal bundles $b' : \nu_{M \# \Sigma} \to \nu_{M'}$ which sends the collapsing map into the collapsing map. But the map which collapses Σ to a point $M \# \Sigma \to M$ is normally cobordant to the identity $M \to M$, so the result follows in one direction.

The other direction follows from (II.3.7). □

Thus the homotopy spheres ∂U which are boundaries of parallelizable manifolds U, play an important role in studying closed manifolds. Removing a disk from interior of U, we see that these are the homotopy spheres which admit normal maps which are normally cobordant to $(1, b)$, $1 : S^m \to S^m$ is the identity. By (II.3.7) if m is even, ∂U is h-cobordant to S^m. If $m = 4k + 1$, since the obstruction σ to making an h-cobordism is in $\mathbb{Z}_2$, by the Addition Theorem $\partial U \# \partial U$ is h-cobordant to S^m. If $m = 4k - 1$, there are parallelizable manifolds W^{4k} with non-zero index and $\partial W = S^m$, (see [37]). Let $N_k = g.c.d$ (index W^{4k}) over such W^{4k}. Now $W - \operatorname{int} D^m$ defines a normal cobordism between S^m and S^m, and for this normal cobordism $\sigma = 1/8$ (index W). Hence $8 \mid \operatorname{index} W$ by (III.3.10), so $8 \mid N_k$. It follows that if index $U = nN$, then $U \amalg (-nW)$ has index 0, and hence $\partial U \# (-nS^m) = \partial U$ is h-cobordant to S^m. If we define bP^{m+1} to be the set of h-cobordism classes of homotopy m-spheres which bound parallelizable manifolds made into a group using the connected sum operation, we may deduce the theorem of Kervaire-Milnor [34]:

II.3.9 Theorem. *$bP^{m+1}=0$ for m even, $m \geqq 4$, cyclic of order at most 2 for $m=4k+1$, cylic of order $N_k/8$ for $m=4k-1$.*

Proof. We have shown above that $bP^{m+1}=0$ for m even, and that $2x=0$ for $x \in bP^{4k+2}$, and $nx=0$, for $x \in bP^{4k}$, $n=N_k/8$. Let (g, c), $g:(U, \partial U) \to (D^{m+1}, S^m)$ be such that $\sigma(g, c)=1$, (using the Plumbing Theorem (II.1.3)). If $\Sigma^m \in bP^{m+1}$, $\Sigma^m = \partial W^{m+1}$, and (f, b) is a normal map $f:(W_0, \Sigma^m \cup S^m) \to (S^m \times I, S^m \times 0 \cup S^m \times 1)$, $W_0 = W - \operatorname{int} D^{m+1}$, then if $\sigma(f, b)=r$, $\sigma((f, b) \amalg (-r)(g, c))=0$ by the Addition Theorem (II.1.4), (where $-$ indicates negative orientation), where the sum is along a disk in $S^m \subset \partial W_0$. Then Σ is h-cobordant to $(-r)(\partial U)$ by (II.1.2). □

More details on the exact order of bP^{m+1}, the group structure, etc., are found in [34].

Now we have the theorem of Wall [65].

II.3.10 Theorem. *Let (X, Y) be a Poincaré pair of dimension $m \geqq 6$, X and Y 1-connected, $Y \neq \emptyset$, and let ξ^k be a k-plane bundle over X, $\alpha \in \pi_{m+k}(T(\xi), T(\xi \mid Y))$ such that $h(\alpha) \cap U = [X] \in H_m(X, Y)$. Then the normal map represented by α is normally cobordant to a homotopy equivalence (f, b), $f:(M, \partial M) \to (X, Y)$, which is unique up to h-cobordism. Hence (X, Y) has the homotopy type of a differentiable manifold, unique up to h-cobordism in the given normal cobordism class.*

Proof. Let $(f', b'):(M', \partial M') \to (X, Y)$ be a normal map representing α. By the Cobordism Property (II.1.5), $\sigma(f' \mid \partial M', b' \mid \partial M')=0$, so by the Fundamental Theorem (II.1.2), $(f' \mid \partial M', b' \mid \partial M')$ is normally cobordant to a homotopy equivalence. This normal cobordism extends to a normal cobordism of (f', b') to (f'', b'') such that $f'' \mid \partial M''$ is a homotopy equivalence (compare with proof of (II.3.4)). By (II.3.6), (f'', b'') is normally cobordant to a homotopy equivalence, (f, b).

Let (f_i, b_i), $i=0, 1$ be two normal maps which are homotopy equivalences, and in the class of α, so (f_0, b_0) is normally cobordant to (f_1, b_1). Let (F, B) be the normal cobordism, $F:(W, V) \to (X \times I, Y \times I)$, $\partial W = M_0 \cup V \cup M_1$, $\partial M_0 \cup \partial M_1 = \partial V$, $F(x)=(f_i(x), i)$ for $x \in M_i \subset W$. This gives a normal map into $(X \times I, X \times 0 \cup Y \times I \cup X \times 1)$ and by (II.3.6), $(F, B) \amalg (g, c)$ is normally cobordant rel$(X \times 0 \cup Y \times I \cup X \times 1)$ to a homotopy equivalence, where $g:(U, \partial U) \to (D^{m+1}, S^m)$. But if (g, c) is added to (F, B) along a disk in V, then M_0 and M_1 and $F \mid M_i = f_i$ remain as they were, so we get an h-cobordism between M_0 and M_1 (or between (f_0, b_0) and (f_1, b_1)). □

Similar to (II.3.8) we obtain

II.3.11 Corollary. *Let M, M' be compact smooth 1-connected manifolds, of dimension $m \geqq 6$, and with ∂M, $\partial M'$ 1-connected and non-empty. Then a homotopy equivalence $f:(M, \partial M) \to (M', \partial M')$ is homotopic* *to a diffeomorphism* $f': M \to M'$, *if and only if there is a linear bundle map* $b: \nu \to \nu'$ *covering* f (ν, ν' *normal bundles of* M, M' *in* D^{m+k}) *such that* $T(b)_*(\alpha) = \alpha'$, (*where* α, α' *are the homotopy classes of the collapsing maps*, $\alpha \in \pi_{m+k}(T(\nu), T(\nu|\partial M))$, $\alpha' \in \pi_{m+k}(T(\nu'), T(\nu'|\partial M'))$).

§ 4. Reinterpretation Using the Spivak Normal Fibre Space

Now we shall reinterpret the results of II § 3 in the terms of the Spivak normal fibre space. In particular we will describe the classification theorem of Sullivan [62] from this point of view, rather than the "dual" approach of [62].

Now we refer to the work of Stasheff [58] or Brown [14], which shows that there is a classifying space, called B_{G_n}, for fibre spaces with a homotopy $(n-1)$-sphere as fibre, in the appropriate category of spaces.

II.4.1 Theorem. (Stasheff). *Consider the category* $\mathscr{C}$ *of spaces with the homotopy type of locally finite CW complexes. Then there is* B_{G_n} *in* $\mathscr{C}$ *and a* $(n-1)$*-spherical fibre space* γ_n *over it such that if* X *is in* $\mathscr{C}$ *and* ξ *is an* $(n-1)$*-spherical fibre space over* X *then there is a map* $f: X \to B_{G_n}$ *such that* $f^*(\gamma_n)$ *is fibre homotopy equivalent to* ξ. *Further, if* $\xi_1 = f_1^* \gamma_n$ *and* $\xi_2 = f_2^* \gamma_n$, $f_i: X \to B_{G_n}$, *and* ξ_1 *is fibre homotopy equivalent to* ξ_2, *then* f_1 *is homotopic to* f_2.

We refer to [58] for the proof.

Let B_G be the classifying space for k-spherical fibre spaces where k is large. Its homotopy type in low dimensions (i.e. $< k-1$) is independent of k, so we suppress k in our notation. This fact is a consequence of the Freudenthal Suspension Theorem, (compare (I.4.10)). Similarly if B_O is the classifying space for $(k+1)$-plane bundles, the homotopy type in low dimensions is independent of k and we omit k in the notation (see [44, 32]). Since the complement of the zero cross-section is a k-sphere bundle, we have a natural map $\varrho: B_O \to B_G$.

II.4.2 Corollary. *Let* X *be a* 1*-connected Poincaré duality space of dimension* $n \geqq 5$, *and let* $f: X \to B_G$ *be the classifying map of its Spivak normal spherical fibre space. If* n *is odd,* X *has the homotopy type of a smooth manifold if and only if* f *factors through* $\varrho: B_O \to B_G$. *If* $n = 4k$, X *has the homotopy type of a smooth manifold if and only if there is a* $g: X \to B_O$ *such that* $\varrho g \sim f$ *and* $\langle L_k(g^*(\omega^{-1})), [X] \rangle = \operatorname{index} X$, *where* ω *is the canonical linear bundle over* B_O.

II.4.3 Corollary. *Let* (X, Y) *be an* n*-dimensional Poincaré pair with* X *and* Y 1*-connected,* $Y \neq \emptyset$ *and* $n \geqq 6$, *and let* $f: X \to B_G$ *be the classifying map of its Spivak normal spherical fibre space. Then* (X, Y) *has the homo-*

topy type of a smooth manifold with boundary if and only if f factors through $\varrho: B_O \to B_G$.

If ν^k is the normal spherical fibre space of X, X a Poincaré duality space of dimension n, or (X, Y) a Poincaré duality pair, then by definition there is an $\alpha \in \pi_{n+k}(T(\nu))$, or $\alpha \in \pi_{n+k}(T(\nu), T(\nu|Y))$ such that $h(\alpha) \cap U = [X]$. Then if $g: X \to B_O$ such that $\varrho g \sim f$, then $(\varrho g)^*(\gamma) = \nu$ and $(\varrho g)^*(\gamma) = g^*(\varrho^* \gamma)$, where $\varrho^*(\gamma) =$ canonical bundle over B_{O_k}. Hence $g^*(\varrho^* \gamma) = \xi$ is a linear k-plane bundle over X which is fibre homotopy equivalent to ν. Hence, there is $\alpha' \in \pi_{n+k}(T(\xi))$ or $\alpha' \in \pi_{n+k}(T(\xi), T(\xi|Y))$ with $h(\alpha') \cap U = [X]$. Then (II.4.2) and (II.4.3) follow from (II.3.1) and (II.3.10). □

In [7, § 4], we studied the general situation of "reducing" the structural group of a bundle, or giving it a "structure" in another theory, which we shall specialize here in the context of spherical fibre spaces and linear structure.

Let ξ be a k-spherical fibre space, over a finite complex X, k very large. A *linear structure* on ξ will be a map of fibre spaces $\alpha: \xi \to \bar{\gamma}$ (of degree 1 on each fibre), where $\bar{\gamma} = \varrho^*(\gamma)$, $\varrho: B_O \to B_G$, γ the canonical k-spherical fibre space over B_G, so that $\bar{\gamma}$ is the canonical bundle over B_O. Two linear structures $\alpha_0, \alpha_1: \xi \to \bar{\gamma}$ are *equivalent* (or *concordant*) if there is a linear structure $A: \xi \times I \to \bar{\gamma}$ such that $A(x, i) = \alpha_i(x)$, $i = 0, 1$, $x \in \xi$.

Let G/O be the fibre of ϱ (if ϱ has been made into a fibre map $\varrho: B_O \to B_G$). One can identify this fibre in a natural way with the orbit space of G by the action of O, where $G = \{f: S^k \to S^k, \deg f = \pm 1\}$ with the compact open topology, $O = O_{k+1}$ is the orthogonal group, k very large.

II.4.4 Theorem. *Equivalence classes of linear structures on ξ (provided there exists one) are in $1-1$ correspondence with $[X, G/O]$, the homotopy classes of maps of X into G/O.*

This is a special case of the situation considered in [7, (4.2)]. We outline the proof in this context, referring to [7] for some of the detailed arguments.

Proof. Let $\beta: \bar{\gamma} \to \gamma$ be the map of fibre spaces covering $\varrho: B_O \to B_G$, and let $\bar{\alpha}: \xi \to \gamma$ be a classifying bundle map for ξ. A *normal linear structure* on ξ will be a linear structure $\alpha: \xi \to \bar{\gamma}$ such that $\beta\alpha = \bar{\alpha}$. An equivalence $A: \xi \times I \to \bar{\gamma}$ will be called normal if $\beta A = \bar{\alpha} p_1$, $p_1: \xi \times I \to \xi$ is projection. Let $\mathscr{S}(\xi)$ $(\mathscr{S}_0(\xi))$ denote the set of equivalence (normal equivalence) classes of linear structures (normal linear structures) on ξ. Clearly there is a natural map $\varepsilon: \mathscr{S}_0(\xi) \to \mathscr{S}(\xi)$.

II.4.5 Lemma. *ε is a $1-1$ correspondence.*

Proof. Using the covering homotopy theorem one can easily show any α is equivalent to α', which is normal, by covering the homotopy of

ϱa to $\bar{a}$ by a homotopy of $\beta\alpha$ to $\bar{\alpha}$, (where $a: X \to B_O$, $\bar{a}: X \to B_G$ are the maps induced on base spaces by α, $\bar{\alpha}$). A similar argument on an equivalence between normal structures, shows that one can find a homotopic normal equivalence (see [7, (4.1)]). □

II.4.6 Lemma. *If $\mathscr{S}_0(\xi)$ is non-empty, then $\mathscr{S}_0(\xi) \cong [X, G/O]$.*

Proof. Let B_{O_k}, B_{G_k} be the classifying spaces of k-plane bundles, $(k-1)$ spherical fibre spaces, and let $B_O = \cup B_{O_k}$, $B_G = \cup B_{G_k}$, be the limit spaces. Let $\varrho_k: B_{O_k} \to B_{G_k}$ be the map inducing the canonical bundle, and let $\varrho: B_O \to B_G$ be the limit map.

Now B_O and B_G are H-spaces with multiplication induced by Whitney sum and $\varrho: B_O \to B_G$ is a multiplicative map. Hence G/O, the fibre, is also an H-space. Also the inclusions induce isomorphisms

$$[X, B_{O_k}] \cong [X, B_O],\ [X, B_{G_k}] \cong [X, B_G],\ [X, G_k/O_k] \cong [X, G/O],$$

for $\dim X < k-1$. Hence we may multiply maps into B_{G_k}, B_{O_k}, $G_k/O_k =$ the fibre of ϱ_k, provided the domain X has dimension $< k-1$.

If $\alpha: \xi \to \bar{\gamma}$ is a representative of an element x in $\mathscr{S}_0(\xi)$, then $\beta\alpha = \bar{\alpha}$, so if α covers $a: X \to B_O$, then $\varrho a = \bar{a}$. Take $\alpha_0: \xi \to \bar{\gamma}$ representing a fixed element $x_0 \in \mathscr{S}_0(\xi)$, α_0 covers a_0.

Now the structures in $\mathscr{S}_0(\xi)$ are in $1-1$ correspondence with homotopy classes of maps $a: X \to B_O$ such that

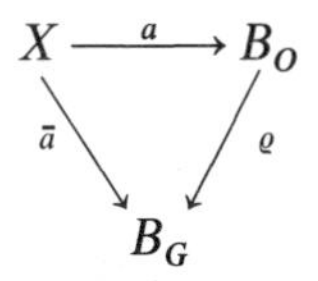

commutes, and homotopies such that

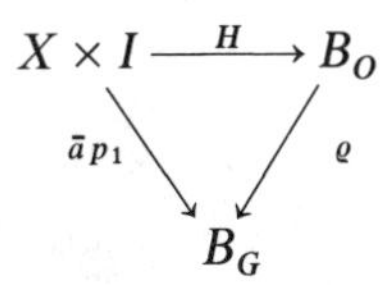

commutes.

For a and $\bar{\alpha}: \xi \to \gamma$ define a map $\alpha: \xi \to \bar{\gamma}$ since $\bar{\gamma}$ is induced from γ by ϱ, and similarly a homotopy defines an equivalence in $\mathscr{S}_0(\xi)$ and vice versa. It follows that elements of $\mathscr{S}_0(\xi)$ are in $1-1$ correspondence with homotopy classes of sections of $E \to X$ where E is induced from the fibre space $B_O \to B_G$ with fibre G/O by the map $a: X \to B_G$. Since G/O acts on B_O so that

$$\begin{array}{ccc} G/O \times B_O & \longrightarrow & B_O \\ \downarrow{\scriptstyle \pi p_2} & & \downarrow{\scriptstyle \pi} \\ B_G & \xrightarrow{1} & B_G \end{array}$$

commutes, G/O acts similarly on E, $\mu: G/O \times E \to E$. If a_0 corresponds to a section $s_0: X \to E$, then $T: G/O \times X \to E$ given by $T(x, y) = \mu(x, s_0(y))$ defines a homotopy equivalence. Then with this representation other sections correspond to maps $X \to G/O \times X$ with component in X being the identity, or in other words maps $X \to G/O$. Similarly homotopy of sections corresponds to homotopy in G/O. □

Applying (II.4.5) and (II.4.6) yields (II.4.4). □

If $\alpha: \xi \to \bar{\gamma}$ is a linear structure on ξ, and if $a: X \to B_O$ is the classifying map on base spaces, then α defines a fibre homotopy equivalence $f: \xi \to a^*(\bar{\gamma})$, ($\bar{\gamma}$ being a linear bundle). An equivalence A between α_0 and α_1 induces a fibre homotopy equivalence $f: \xi \times I \to A^*(\bar{\gamma})$ extending f_0, f_1 induced on $\xi \times 0$ and $\xi \times 1$ by α_0 and α_1. Since $A^*(\bar{\gamma})$ is a bundle over $X \times I$, there is a linear equivalence $B: A^*(\bar{\gamma}) \to a_0^*(\bar{\gamma}) \times I$ extending the identity on $a_0^*(\bar{\gamma})$. Hence $b: a_1^*(\bar{\gamma}) \to a_0^*(\bar{\gamma}) \times 1$, $b = B \mid a_1^*(\bar{\gamma})$ is a linear equivalence, and $b f_1$ is fibre homotopic to f_0.

Now consider pairs (η, α) where η is a linear k-plane bundle over X, $\alpha: \xi \to \eta$ is a fibre homotopy equivalence covering the identity of X. Call such a pair (η, α) a G/O bundle (structure on ξ). Two G/O bundles (η_i, α_i), $i = 0, 1$ are equivalent if there is a G/O bundle $(\bar{\eta}, \bar{\alpha})$ over $X \times I$, $\bar{\alpha}: \xi \times I \to \bar{\eta}$, and linear equivalences $b_i: \bar{\eta} \mid X \times i \to \eta_i$ such that $b_i(\bar{\alpha} \mid \xi \times i) = \alpha_i$, $i = 0, 1$. This is equivalent to the statement that there exists a linear equivalence $b: \eta_0 \to \eta_1$ such that $b\alpha_0$ is fibre homotopic to α_1. Thus we get

II.4.7 Proposition. *Equivalence classes of linear structures on ξ are in $1-1$ correspondence with equivalence classes of G/O bundle structures on ξ.* □

Now let ξ be the Spivak normal fibre space of a Poincaré pair (X, Y), (see I § 4).

II.4.8 Lemma. *A normal map (f, b), $f: (M, \partial M) \to (X, Y)$, $b: \nu \to \eta$, η a linear k-plane bundle, determines a linear structure on ξ, depending only on the normal cobordism class of (f, b). Two normal maps (f_i, b_i), $i = 1, 2$ determine equivalent linear structures if and only if there is a linear bundle equivalence $b': \eta_1 \to \eta_2$ such that (f_2, b_2) is normally cobordant to $(f_1, b' b_1)$.*

Proof. By (I.4.19) there is a fibre homotopy equivalence $h: \xi \to \eta$ such that $T(h)_* (\delta_0) = T(b)_* (\alpha)$, where $\delta_0 \in \pi_{m+k}(T(\xi), T(\xi \mid Y))$ is a fixed element such that $h(\delta_0) \cap U = [X]$, and $\alpha \in \pi_{m+k}(T(\nu), T(\nu \mid \partial M))$ is the homotopy class of the collapsing map. By (I.4.19), such an h is unique up to fibre homotopy, so this defines a map φ from the set of normal maps $\mathcal{N}$ into G/O bundle structures on ξ. If (f_i, b_i), $i = 1, 2$ are normally cobordant, then $\eta_1 = \eta_2 = \eta$ and $T(b_1)_* (\alpha_1) = T(b_2)_* (\alpha_2)$ by (II.2.14). It follows then that the corresponding structures $h_i: \xi \to \eta$ are homotopic, so the map

$\varphi:\mathcal{N}\to\mathcal{S}(\xi)$ depends only on the normal cobordism class. Hence φ defines $\varphi_0:\mathcal{N}_0\to\mathcal{S}(\xi)$, where $\mathcal{N}_0=$ set of normal cobordism classes. If $\varphi_0(f_1,b_1)=\varphi_0(f_2,b_2)$, then $h_2=b'h_1$ where $b':\eta_1\to\eta_2$. Then

$$T(b')_*\,T(h_1)_*(\delta_0)=T(b')_*\,T(b_1)_*(\alpha_1)=T(b'b_1)_*(\alpha_1)=T(h_2)_*(\delta_0)=T(b_2)_*(\alpha_2)\,.$$

Hence (f_2,b_2) is normally cobordant to $(f_1,b'b_1)$. □

Putting together (II.4.8) with (II.3.7) and (II.3.10), and with (II.4.4) we obtain the theorem of Sullivan [62]:

Let (X,Y) be a Poincaré pair. Define $\mathcal{S}(X)$ to be the set of pairs (h,M) where M is a smooth manifold with boundary, $h:(M,\partial M)\to(X,Y)$ is a homotopy equivalence of pairs, under the equivalence relation $(h_0,M_0)\sim(h_1,M_1)$ if there is an h-cobordism W^{m+1}, and a map $H:(W,V)\to(X,Y)$, $(\partial W=M_0\cup V\cup M_1)$ such that $H|M_i=h_i$, $i=0,1$.

II.4.9 Theorem. *Let $(M,\partial M)$ be a compact smooth manifold with boundary, dimension $M\geqq 6$, M and ∂M 1-connected, $\partial M\neq\emptyset$. Then $\mathcal{S}(M)$ is in $1-1$ correspondence with $[M,G/O]$.*

In case $\partial M=\emptyset$, the analogous theorem holds modulo homotopy spheres which bound π-manifolds, (compare (II.3.7) and (II.3.10)), but in this case the natural expression in terms of an *exact sequence* (which has a generalization to the non-simply connected case): Let M be a closed smooth 1-connected manifold of dimension $m\geqq 5$.

II.4.10 Exact sequence of surgery. *There is an exact sequence of sets*

$$P_{m+1}\xrightarrow{\ \omega\ }\mathcal{S}(M)\xrightarrow{\ \eta\ }[M,G/O]\xrightarrow{\ \sigma\ }P_m$$

where

$$P_i=\begin{cases}0 & i\ odd\\ \mathbb{Z} & i=4k\\ \mathbb{Z}_2 & i=4k+2\end{cases}$$

where η is defined by the normal cobordism class, σ is the surgery obstruction of the normal map, and $\omega(x)$ is defined as below.

Taking connected sum along the boundary of $M\times[0,1]$ and V^{m+1} where (g,c) is a normal map $g:V\to D^{m+1}$, $g|\partial V$ a homotopy equivalence, $\sigma(g,c)=x$, we obtain a manifold with boundary $=M\cup M\#\partial V$. Define $\omega(x)$ to be $M\#\partial V$ with the obvious homotopy equivalence which collapses $\partial V-$ cell to a point. This actually defined an *action* of P_{m+1} on $\mathcal{S}(M)$ as follows: If $h:M'\to M$ represents $y\in\mathcal{S}(M)$, and $x\in P_{m+1}$, then $\omega'(x)\in\mathcal{S}(M')$ is defined as above. Let $\omega(x,y)=h'_*(\omega'(x))$, i.e. if (M'',h'), $h':M''\to M'$ represents $\omega'(x)$, then (M'',hh') represents $\omega(x,y)\in\mathcal{S}(M)$. Then the sequence of (II.4.10) is exact in the stronger sense.

II.4.11. *$\eta(y)=\eta(y')$, $y, y' \in \mathscr{S}(M)$, if and only if $y'=\omega(x,y)$ for some $x \in P_{m+1}$.*

The piecewise linear version of (II.4.9) may be proved in a similar way using p.l. microbundle theory, and surgery on p.l. manifolds (see [48, 13]) and is an important step in the proof of the Hauptvermutung for 1-connected manifolds M, with ∂M 1-connected, dimension $M \geqq 6$, and $H_3(M)$ having no 2-torsion (see [63]). (II.4.10) and (II.4.11) degenerate in the p.l. case, but become interesting again in their non-simply connected versions (see [66]).

III. The Invariant σ

In this chapter we prove the Invariant Theorem of Chapter II, § 1 and deduce the three properties of σ of (II, § 1).

We will outline a slightly more general version than that indicated in Chapter II. Let (X, Y) be an m-dimensional oriented Poincaré duality pair, i.e., there is an element $[X]$ (the orientation class) in $H_m(X, Y)$ such that

$$[X]\cap : H^q(X) \to H_{m-q}(X, Y)$$

is an isomorphism for all q.

Recall that in Chapter I we showed that $[X]\cap : H^q(X) \to H_{m-q}(X, Y)$ being an isomorphism for all q is equivalent to $[X]\cap : H^q(X, Y) \to H_{m-q}(X)$ being an isomorphism for all q, and that this implies that in the diagram

$$\begin{array}{ccccccccc}
\cdots \to H^q(X, Y) & \xrightarrow{j^*} & H^q(X) & \xrightarrow{i^*} & H^q(Y) & \xrightarrow{\delta} & H^{q+1}(X, Y) \to \cdots \\
\downarrow {\scriptstyle [X]\cap} & & \downarrow {\scriptstyle [X]\cap} & & \downarrow {\scriptstyle [Y]\cap} & & \downarrow {\scriptstyle [X]\cap} \\
\cdots \to H_{m-q}(X) & \xrightarrow{j_*} & H_{m-q}(X, Y) & \xrightarrow{\partial} & H_{m-q-1}(Y) & \xrightarrow{i_*} & H_{m-q-1}(X) \to \cdots
\end{array}$$

(where $[Y] = \partial[X] \in H_{m-1}(Y)$, $i: Y \to X$, $j: X \to (X, Y)$ are inclusions) all the vertical arrows are isomorphisms. In particular Y is an oriented Poincaré duality space of dimension $m-1$.

Let $m = 4k$, and let $f: (X_1, Y_1) \to (X_2, Y_2)$ be a map of degree 1 of the Poincaré duality pairs (X_i, Y_i), $i = 1, 2$ such that $(f|Y_1)_*: H_*(Y_1) \to H_*(Y_2)$ is an isomorphism.

A cobordism of f rel Y_2 is described by U with subsets

$$X_1, X_1', Y_1 = X_1 \cap X_1',$$

such that $(U, X_1 \cup X_1')$ is an $(m+1)$-dimensional Poincaré pair, with orientation $[U]$ compatible with the orientation $[X_1]$ (see I, § 2) and a map $F: (U, Y_1) \to (X_2, Y_2)$, such that $F|(X_1, Y_1) = f$. We write below $A = X_1 \cup X_1'$. Then $I(f) \in \mathbb{Z}$ is defined in § 2 such that if f is cobordant rel Y_2 to a homology isomorphism, then $I(f) = 0$.

Let $f:(X_1, Y_1)\to(X_2, Y_2)$ be as above and let v_i be a $(k-1)$- spherical fibre space over (X_i, Y_i), k large, and $b: v_1\to v_2$ a map of fibre spaces covering f. Suppose that $\alpha\in\pi_{m+k}(T(v_1))$ is such that $h(\alpha)\cap U_1=[X_1]$, so that v_1 is the normal spherical fibre space of Spivak for (X_1, Y_1), and it follows that $h(T(b)_*(\alpha))\cap U_2=[X_2]$, so that v_2 is the normal spherical fibre space of (X_2, Y_2) (see I, § 4 and [57]). Then the pair (f, b) is called a normal map of the Poincaré pairs, (compare II, § 1). A normal cobordism of (f, b) is a cobordism rel Y_2 of f, as above, and in addition a $(k-1)$-spherical fibre space $\bar{v}$ over U, a map of fibre spaces $\bar{b}:\bar{v}\to v_2$ and an element $\bar{\alpha}\in\pi_{m+1+k}(T(\bar{v}), T(\bar{v}|A))$ such that $T(\bar{b}|(\bar{v}|(X_1, Y_1))_*(\partial\bar{\alpha})=T(b)_*(\alpha)$ where $\partial:\pi_{m+1+k}(T(\bar{v}), T(\bar{v}|A))\to\pi_{m+k}(T(\bar{v}|X_1), T(\bar{v}|Y_1))$ is the natural boundary (again compare II, § 1).

If $m=4k$, and (f, b) is a normal map, then $I(f)$ is divisible by 8 and we define $\sigma(f, b)=\frac{1}{8}I(f)$. If $m=4k+2$, we define $\sigma(f, b)\in\mathbb{Z}_2$, such that if (f, b) is normally cobordant to a homology isomorphism with $\mathbb{Z}_2$ coefficients, then $\sigma(f, b)=0$.

We will also deduce the various properties of σ needed.

§ 1. Quadratic Forms over $\mathbb{Z}$ and $\mathbb{Z}_2$

Let V be a finitely generated free $\mathbb{Z}$-module, and let $(\,,\,)$ be a symmetric bilinear form on V so that

(i) $(x, y)=(y, x)$,

(ii) $(\lambda x+\lambda' x', y)=\lambda(x, y)+\lambda'(x', y)$, $\lambda, \lambda'\in\mathbb{Z}$, $x, x', y\in V$.

Choosing a basis $\{b_i\}$ for V, $i=1,\dots,n$, and letting $a_{ij}=(b_i, b_j)$ represent $(\,,\,)$ as a matrix $A=(a_{ij})$, and $(x, y)=xAy^t$ in terms of this basis, where $x=\sum_{i=1}^{n}\lambda_i b_i$, etc. (^{t} means transpose). If we change the basis by an invertible $n\times n$ matrix M so that $b'=Mb$, i.e., $b'_i=\Sigma_j m_{ij}b_j$, then in terms of the new basis, $(\,,\,)$ is represented by the matrix MAM^t. Such changes are equivalent to doing a sequence of row and column operations on A, performing the same operation on row and column. For example we may add λ(i-th row) to the j-th row, and then λ(i-th column) to the j-th column.

The bilinear form $(\,,\,)$ defines $q: V\to\mathbb{Z}$ by $q(x)=(x, x)$. Then $(x, y)=\frac{1}{2}(q(x+y)-q(x)-q(y))$ so that $(\,,\,)$ is derivable from q. Then $(\,,\,)$ is called the associated bilinear form to the quadratic form q.

The bilinear $(\,,\,)$ defines naturally a bilinear form (again denoted by $(\,,\,)$) on $V\otimes\mathbb{Q}$ into $\mathbb{Q}$.

III.1.1 Proposition. *If $(\,,\,)$ is a symmetric bilinear form on a finite dimensional vector space V' over $\mathbb{Q}$ into $\mathbb{Q}$, then there is a basis for V' such that the matrix of $(\,,\,)$ is diagonal.*

The proof is a routine exercise.

Now we may define the index or signature of $(\,,\,)$ to be the number of positive entries on the diagonal minus the number of negative entries (in the diagonalized matrix). The first number is the dimension of the maximal subspace on which $(\,,\,)$ is positive definite, (i.e., $(x, x) > 0$, if $x \neq 0$) and the second is the dimension of the maximal subspace on which $(\,,\,)$ is negative definite (i.e., $(x, x) < 0$ if $x \neq 0$). It follows that the signature is an invariant, i.e., it does not depend on the choice of basis. Hence we have defined an invariant

$$\operatorname{sgn}: (\text{Quadratic forms over } \mathbb{Z}) \to \mathbb{Z}.$$

We shall call a quadratic form over $\mathbb{Z}$ *non-singular* if the determinant $|A| = \pm 1$ (i.e., if $(\,,\,)$ is unimodular). Over a field we call it non-singular if $|A| \neq 0$.

III.1.2 Proposition. *Let q be a non-singular quadratic form on a finite dimensional V over $\mathbb{R}$, the reals. Then $\operatorname{sgn}(q) = 0$ if and only if there is a subspace $U \subset V$ such that*

(i) $\dim_R U = \frac{1}{2} \dim_R V$

(ii) $(x, y) = 0$ *for* $x, y \in U$.

Proof. Let V_+ and V_- be subspaces of V such that q is positive definite on V_+, negative definite on V_-, and V_+, V_- are maximal with respect to this property. Then $\operatorname{sgn}(q) = \dim V_+ - \dim V_-$. Clearly $V_+ \cap V_- = 0$ and since q is non-singular, $V = V_+ + V_-$. Now

$$V_+ \cap U = V_- \cap U = 0$$

since $(\,,\,)$ is zero on U. On the other hand

$$\dim(V_+ \cap U) \geqq \dim V_+ + \dim U - \dim V$$

$$\dim(V_- \cap U) \geqq \dim V_- + \dim U - \dim V$$

so that $\dim V_\pm = \dim V - \dim U = \frac{1}{2} \dim V$, and thus $\operatorname{sgn}(q) = 0$.

If $\operatorname{sgn}(q) = 0$, then $\dim V_+ = \dim V_-$. Over the reals $\mathbb{R}$, one may find orthonormal bases for V_+ and V_-, $\{a_i\}, \{b_i\}$ respectively, $i = 1, \ldots, n$, such that $(a_i, a_j) = \delta_{ij}$, $(b_i, b_j) = -\delta_{ij}$, $(a_i, b_j) = 0$. Then $c_i = a_i + b_i$, $i = 1, \ldots, n$, generates U, and $(c_i, c_i) = (a_i, a_i) + (b_i, b_i) = 1 - 1 = 0$, and $(c_i, c_j) = 0$, so $q = 0$ on U. □

Now we state some non-trivial results which we will need, which we shall not prove here.

III.1.3 Proposition. *Let q be a non-singular quadratic form $q: V \to \mathbb{Z}$, and suppose q is indefinite (i.e., neither positive definite nor negative definite). Then there is $x \in V$, $x \neq 0$ such that $q(x) = 0$.*

For a proof see [45, Lemma 8] (see also [46]).

III.1.4 Proposition. *Let q be a non-singular quadratic form $q: V \to \mathbb{Z}$ and suppose $2|q(x,x)$ for all $x \in V$ (we say q is an even form). Then $8|\mathrm{sgn}(q)$.*

For a proof see [46].

Now we consider the field $\mathbb{Z}_2$ and consider a function $q: V \to \mathbb{Z}_2$, where V is a $\mathbb{Z}_2$ vector space of finite dimension over $\mathbb{Z}_2$. We shall call q a quadratic form if $q(0)=0$ and

$$q(x+y)-q(x)-q(y)=(x,y)$$

is bilinear. It is clear that $(x,y)=(y,x)$ and $(x,x)=q(2x)-2q(x)=0$ so that $(\,,\,)$ is a symplectic bilinear form. Thus if $(\,,\,)$ is non-singular we may find a basis $a_i, b_i, i=1,\ldots,n$ for V such that $(a_i,b_j)=\delta_{ij}, (a_i,a_j)=(b_i,b_j)=0$ (see [3]). Thus in case q(i.e. $(\,,\,)$) is non-singular with respect to the symplectic basis $\{a_i,b_i\}$ we define the *Arf invariant* (see [2]):

$$c(q)=\sum_{i=1}^{n} q(a_i)q(b_i) \in \mathbb{Z}_2 .$$

We shall show that c is independent of the choice of base, and completely determines q up to equivalence.

First we consider the 2-dimensional vector space U, with basis $a, b, (a,b)=1, (a,a)=(b,b)=0$. There are two quadratic forms on U compatible with $(\,,\,), q_i: U \to \mathbb{Z}_2, i=0,1, q_1(a)=q_1(b)=1$, and $q_0(a)=q_0(b)=0$. Note that $q_1(a+b)=q_0(a+b)=1$.

III.1.5 Lemma. *Any non-singular quadratic form on a 2-dimensional space U is isomorphic to either q_0 or q_1.*

The proof is trivial.

Obviously q_0 is not equivalent to q_1. Also $c(q_0)=0$ and $c(q_1)=1$. Hence the Arf invariant c characterizes non-singular quadratic forms in dimension 2.

III.1.6 Lemma. *On $U+U$, q_0+q_0 is isomorphic to q_1+q_1.*

Proof. Let a_i, b_i $i=1,2$ be a basis for $U+U$ so that a_i, b_i forms a symplectic basis of the i-th U, and if $\psi_i=q_i+q_i$, $i=0,1$ on $U+U$, then $\psi_0(a_i)=\psi_0(b_i)=0$, $i=1,2$, and $\psi_1(a_i)=\psi_1(b_i)=1$, $i=1,2$. Choose a new basis for $U+U$,

$$a_1'=a_1+a_2, \qquad b_1'=b_1+a_2,$$

$$a_2'=a_2+b_2+a_1+b_1, \quad b_2'=b_2+a_1+b_1 .$$

One checks easily that this defines a symplectic basis and

$$\psi_1(a_i')=\psi_0(a_i), \quad \psi_1(b_i')=\psi_0(b_i)$$

so that ψ_1 is isomorphic to ψ_0. □

III.1.7 Proposition. *Let $q: V \to \mathbb{Z}_2$ be a non-singular quadratic form over $\mathbb{Z}_2$. Then q is equivalent to $q_1 + (m-1)q_0$ if with respect to some basis $c(q) = 1$, and q is equivalent to mq_0 if $c(q) = 0$* ($\dim V = 2m$).

Proof. If $a_i, b_i, i = 1, \ldots, n$ is a symplectic basis for V and if $V_i =$ space spanned by a_i, b_i, let $\psi_i = q \mid V_i$. It is evident that $q = \sum_{i=1}^{n} \psi_i$, and by (III.1.5), ψ_i is equivalent to either q_0 or q_1. By (III.1.6), $2q_1 = 2q_0$, so q is equivalent to either mq_0 or $q_1 + (m-1)q_0$. But $c(q_1 + (m-1)q_0) = 1$ and $c(mq_0) = 0$, which implies the results. □

To complete the study of non-singular quadratic forms over $\mathbb{Z}_2$, it remains to show that $\varphi_1 = q_1 + (m-1)q_0$ and $\varphi_0 = mq_0$ are not equivalent. We prove this by the following

III.1.8 Proposition. *The quadratic form φ_1 sends a majority of elements of V to $1 \in \mathbb{Z}_2$, while φ_0 sends a majority of elements to $0 \in \mathbb{Z}_2$.*

III.1.9 Corollary. *If q is a non-singular quadratic form, then $c(q) = 1$ if and only if q sends a majority of elements to $1 \in \mathbb{Z}_2$.*

Proof of (III.1.8). We proceed by induction, the case of $m = 1$ being trivial.

Let $p(\varphi) =$ number of elements $x \in V$ such that $\varphi(x) = 1$ and let $n(g) =$ number of $x \in V$ such that $\varphi(x) = 0$. Hence $p(\varphi) + n(\varphi) = 2^{2m} =$ number of elements in V (including 0).

III.1.10 Lemma. $p(\varphi + q_0) = 3p(\varphi) + n(\varphi)$, $n(\varphi + q_0) = 3n(\varphi) + p(\varphi)$.

Proof. Any element in $V + U$ is of the form (x, u), $x \in V, u \in U$ and $(\varphi + q_0)(x, u) = \varphi(x) + q_0(u)$. Three of the four elements in U have $q_0 = 0$ and only one has $q_0 = 1$, so for each element $x \in V$ such that $\varphi(x) = 1$ we have three elements (x, u) such that $q_0(u) = 0$ and thus $(\varphi + q_0)(x, u) = 1$. Similarly for each $y \in V$ such that $\varphi(y) = 0$ there is one element $(y, v) \in V + U$ with $q_0(v) = 1$ so $(\varphi + q_0)(y, v) = 1$. Hence $p(\varphi + q_0) = 3p(\varphi) + n(\varphi)$, and the other formula follows similarly.

III.1.11 Corollary. *Set $r(\varphi) = p(\varphi) - n(\varphi)$. Then $r(\varphi + q_0) = 2r(\varphi)$, so that if $r(\varphi) > 0$ then $r(\varphi + q_0) > 0$, and if $r(\varphi) < 0$ then $r(\varphi + q_0) < 0$.*

The proof is immediate.

It follows that since $r(q_1) = 2, r(q_0) = -2$, that $r(q_1 + (m-1)q_0) > 0$, $r(mq_0) < 0$, which proves (III.1.8). Since r is obviously an invariant, it follows that $q_1 + (m-1)q_0$ is not equivalent to mq_0. Thus we have proved:

III.1.12 Theorem. (Arf). *Two non-singular quadratic forms on a $\mathbb{Z}_2$ vector space V of finite dimension are equivalent if and only if they have the same Arf invariant.*

Analogous to (III.1.2) we have:

III.1.13 Proposition. *Let q be a non-singular quadratic form $q: V \to \mathbb{Z}_2$. Then the Arf invariant $c(q)=0$ if and only if there is a subspace $U \subset V$, such that*

(i) *$\operatorname{rank}_{\mathbb{Z}_2} U = \frac{1}{2}\operatorname{rank}_{\mathbb{Z}_2} V$,*
(ii) *$q(x)=0$, all $x \in U$.*

Proof. Let $x, y \in U$, U with properties (i), (ii). Then

$$(x, y) = q(x+y) - q(x) - q(y) = 0$$

since $x, y, x+y \in U$. Hence U is an isotropic subspace, (i.e. $(x,y)=0$, $x, y \in U$) and thus a base $a_1, \ldots, a_m$ for U may be extended to a symplectic basis for U $a_1, b_1, \ldots, a_m, b_m$ (since $(\,,)$ is non-singular). It follows that $c(q) = \sum_{i=1}^{m} q(a_i) q(b_i) = 0$.

Conversely if $c(q)=0$, by (III.1.12) q is equivalent to mq_0, so that $U =$ space spanned by $a_1, \ldots, a_m$ (where a_i, b_i are a base for the 2-dimensional space of q_0) has properties (i) and (ii). □

For a bilinear form $(\,,)$ on a vector space V, we let $R =$ radical of $V = \{x \in V$ such that $(x, y) = 0$ all $y \in V\}$.

If $q: V \to \mathbb{Z}_2$ is a quadratic form with $(\,,)$ as associated bilinear form, then we have defined $c(q)$ only if $R =$ radical of V is zero. But if $q \mid R \equiv 0$, then it is easy to see that q defines q' on V/R and the radical of V/R is zero. In that case we may define $c(q) = c(q')$. However if $q \mid R \not\equiv 0$, then it is easy to see that the Arf invariant does not make sense and in fact the equivalence class of the form is determined by rank V and rank R. Note that in this case $r(\varphi) = 0$.

Thus we have proved:

III.1.14 Theorem. *Let $q: V \to \mathbb{Z}_2$ be a quadratic form over $\mathbb{Z}_2$, $R =$ radical of the associated bilinear form. Then the Arf invariant $c(q)$ is defined if and only if $q \mid R \equiv 0$. In general if $q \mid R \equiv 0$, q is determined up to isomorphism by rank V, rank R and $c(q)$, while if $q \mid R \not\equiv 0$, then q is determined by rank V and rank R.*

§ 2. The Invariant $I(f)$

Let (X, Y), (A, B) be oriented Poincaré pairs of dimension m, let $f: (X, Y) \to (A, B)$ be a map of degree 1, i.e., $f_*[X] = [A]$, $[X] \in H_m(X, Y)$, $[A] \in H_m(A, B)$ the orientation classes. Then as in Chapter I, § 2 we have groups $K^q(X, Y)$, $K^q(X)$, $K^q(Y)$ defined with any coefficient group such that the diagram below is commutative with exact rows and columns:

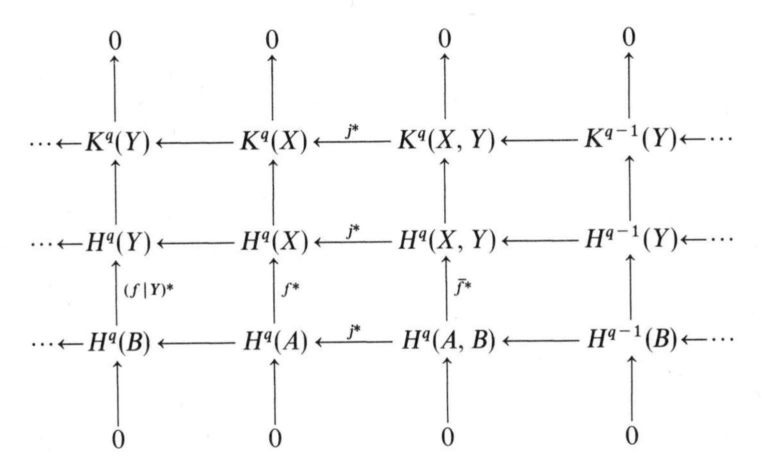

Here the notation $f: X \to A$, $\bar{f}: (X, Y) \to (A, B)$ is used to distinguish the induced cohomology maps. By (I.2.5), the vertical sequences split, $\alpha^*: H^*(X, Y) \to H^*(A, B)$ such that $\alpha^* \bar{f}^* = \text{identity}$, for example, and $K^*(X, Y) = \ker \alpha^*$ by definition.

Suppose $m = \dim(X, Y) = 4k$, and consider the pairing

$$K^{2k}(X, Y; \mathbb{Q}) \otimes K^{2k}(X, Y; \mathbb{Q}) \to \mathbb{Q}$$

defined by $(x, y) = (x \cup y)[X]$. This is symmetric since the dimension is even. Define

III.2.1. $I(f) = \text{signature of } (\,,\,) \text{ on } K^{2k}(X, Y; \mathbb{Q})$.

We note that $(\,,\,)$ is the rational form of the integral form defined on $K^{2k}(X, Y)/\text{torsion}$ by the same formula. If $(f|Y)^*: H^*(B; \mathbb{Q}) \to H^*(Y; \mathbb{Q})$ is an isomorphism, then

$$(x \cup y)[X] = ((j^* x) \cup y)[X], \; j^*: K^{2k}(X, Y; \mathbb{Q}) \to K^{2k}(X; \mathbb{Q})$$

is an isomorphism, and thus from (I.2.9) it follows that $(\,,\,)$ is non-singular. Similarly if $(f|Y)^*: H^*(B) \to H^*(Y)$ is an isomorphism, then the integral form is non-singular. In particular this is of course the case if $Y = B = \emptyset$. We note also that positive degree would have sufficed to define $I(f)$.

III.2.2 Proposition. *Let $f: (X, Y) \to (A, B)$ be a map of degree 1 of Poincaré pairs of dimension $m = 2q + 1$, let F be a field, and consider $i^*: K^q(X; F) \to K^q(Y; F)$ induced by inclusion $i: Y \to X$. Then*

$$\text{rank}_F (\text{image } i^*)^q = \tfrac{1}{2} \text{rank}_F K^q(Y; F).$$

Proof. By (I.2.7) we have a diagram which commutes up to sign:

$$\begin{array}{ccccccc}
\longrightarrow K^q(X;F) & \xrightarrow{i^*} & K^q(Y;F) & \xrightarrow{\delta} & K^{q+1}(X,Y;F) \rightarrow \\
\downarrow{\scriptstyle [X]\cap} & & \downarrow{\scriptstyle [Y]\cap} & & \downarrow{\scriptstyle [X]\cap} \\
\rightarrow K_{q+1}(X,Y;F) & \xrightarrow{\partial} & K_q(Y;F) & \xrightarrow{i_*} & K_q(X;F) \longrightarrow
\end{array}$$

In this diagram the rows are exact and the vertical maps isomorphisms. Hence $(\text{image}\, i^*)^q \cong (\ker i_*)_q$. By (I.2.8), since F is a field,

$$K^q(Y;F) = \text{Hom}(K_q(Y;F), F)$$
$$K^q(X;F) = \text{Hom}(K_q(X;F), F)$$

and $i^* = \text{Hom}(i_*, 1)$. Hence $\text{rank}_F(\text{image}\, i^*)^q = \text{rank}_F(\text{image}\, i_*)_q$ and $\text{rank}_F(\text{image}\, i_*)_q + \text{rank}_F(\ker i_*)_q = \text{rank}_F K_q(Y;F) = \text{rank}_F K^q(Y;F)$. Hence $\text{rank}_F(\text{image}\, i^*)^q = \frac{1}{2}\text{rank}_F K^q(Y;F)$. □

III.2.3 Lemma. *With the hypotheses of* (III.2.2), $(\text{image}\, i^*)^q \subset K^q(Y;F)$ *annihilates itself under the pairing* $(\ ,\ )$.

Proof.

$$(i^*x, i^*y) = ((i^*x) \cup (i^*y))[Y] = (i^*(x \cup y))[Y] = (x \cup y)(i_*[Y]) = 0$$

since $i_*[Y] = i_*\partial[X] = 0$ in $H_{2q}(X)$. □

III.2.4 Theorem. *Let* $f:(X,Y) \to (A,B)$ *be a map of degree 1 of Poincaré pairs of dimension* $m = 4k+1$. *Then* $I(f|Y) = 0$.

Proof. By (III.2.2) $(\text{image}\, i^*)^{2k} \subset K^{2k}(Y;\mathbb{Q})$ is a subspace of $\text{rank} = \frac{1}{2}\text{rank}\, K^{2k}(Y;\mathbb{Q})$ and by (III.2.3) it annihilates itself under the pairing. Hence by (III.1.2), $\text{sgn}(\ ,\ ) = 0$ on $K^{2k}(Y;\mathbb{Q})$, i.e. $I(f|Y) = 0$. □

Now using the notion of sum of Poincaré pairs introduced in Chapter I, § 3, we may study the behavior of I on sums.

Let (X, Y) and (A, B) be Poincaré pairs of dimension m, and suppose each is the sum of pairs, i.e., $X = X_1 \cup X_2$, $X_0 = X_1 \cap X_2$, $Y_i = Y \cap X_i$, $i = 1, 2$, $A = A_1 \cup A_2$, etc., where (X_0, Y_0) and (A_0, B_0) are Poincaré pairs with orientations $\partial_0[X]$, $\partial_0[A]$ respectively (see (I.3.2)). Let $f:(X,Y) \to (A,B)$ be a map of degree 1 such that $f(X_i) \subset A_i$, $i = 1, 2$ (c.f. (I.3.3)). Let $f_i = f|X_i$, $i = 0, 1, 2$.

III.2.5 Theorem. *Suppose* $f:(X,Y) \to (A,B)$ *as above is the sum of two maps* $f_i:(X_i, X_0 \cup Y_i) \to (A_i, A_0 \cup B_i)$, $i = 1, 2$, *and suppose that the map on the intersection* $f_0^*: H^*(A_0, B_0;\mathbb{Q}) \to H^*(X_0, Y_0;\mathbb{Q})$ *is an isomorphism. Then*

$$I(f) = I(f_1) + I(f_2).$$

Proof. Consider the exact sequence induced from the map of Mayer-Viettoris sequences (or the map of triples $f:(X, X_0, Y)\to(A, A_0, B)$) defined by f:

$$\cdots\to K^{q-1}(X_0, Y_0)\to K^q(X_1, X_0\cup Y_1)+K^q(X_2, X_0\cup Y_2)$$
$$\to K^q(X, Y)\to K^q(X_0, Y_0)\to\cdots$$

Since f_0^* is an isomorphism with coefficients $\mathbb{Q}$, it follows that $K^*(X_0, Y_0; \mathbb{Q})=0$, and

III.2.6. $K^q(X, Y; \mathbb{Q}) \cong K^q(X_1, X_0\cup Y_1; \mathbb{Q}) + K^q(X_2, X_0\cup Y_2; \mathbb{Q})$.

Now the map $H^q(X_1, X_0\cup Y_1)+H^q(X_2, X_0\cup Y_2)\to H^q(X, Y)$ is the sum of two maps $H^q(X_i, X_0\cup Y_i)\xleftarrow{\cong} H^q(X, X_{s(i)}\cup Y)\xrightarrow{j_i^*} H^q(X, Y)$ (where $s(i)\neq i, s(i)=1$ or 2). But $j_1^*(x_1)\cup j_2^*(x_2)\in \operatorname{image} H^{2q}(X, (X_1\cup Y)\cup(X_2\cup Y))$ using the relative cup product between $H^q(X, X_1\cup Y)$ and $H^q(X, X_2\cup Y)$. Since $X_1\cup X_2=X$, $H^*(X, (X_1\cup Y)\cup(X_2\cup Y))=0$, so $j_1^*(x_1)\cup j_2^*(x_2)=0$ and it follows that (III.2.6) is an orthogonal decomposition. However the bilinear form restricted to each fator is the usual form on $K^q(X_i, X_0\cup Y_i)$, so the bilinear form on $K^q(X, Y; \mathbb{Q})$ is the sum, and hence $I(f)=I(f_1)+I(f_2)$. □

If (X, Y) is a Poincaré pair of dimension $m=4k$ we may consider the symmetric pairing

$$H^{2k}(X, Y; \mathbb{Q})\otimes H^{2k}(X, Y; \mathbb{Q})\to\mathbb{Q}$$

given by $(x, y)=(x\cup y)[X]$.

III.2.7 Lemma. *$K^{2k}(X, Y; \mathbb{Q})$ and $f^*(H^{2k}(A, B; \mathbb{Q}))$ are orthogonal under the pairing.*

Proof. $(x, y)=(x\cup y)[X]=((j^*x)\cup y)[X]$ where $j: X\to(X, Y)$. But $j^*K^{2k}(X, Y)\subset K^{2k}(X)$, and by (I.2.9), $K^{2k}(X)$ is orthogonal to $f^*H^{2k}(A, B)$, so the lemma follows. □

Thus we may define $I(X, Y)=$ signature of $(\,,)$ on $H^{2k}(X, Y; \mathbb{Q})$.

III.2.8 Theorem. $I(f)=I(X, Y)-I(A, B)$.

Proof. By (III.2.7), $H^{2k}(X, Y; \mathbb{Q})=K^{2k}(X, Y; \mathbb{Q})+f^*H^{2k}(A, B; \mathbb{Q})$ as an orthogonal direct sum, so that the inner product is the sum of those on the factors. But the inner product on $f^*H^{2k}(A, B; \mathbb{Q})$ is the same as that on $H^{2k}(A, B; \mathbb{Q})$ so that it follows that $I(X, Y)=I(f)+I(A, B)$ and the result follows. □

III.2.9 Theorem. *Let $f:(X, Y)\to(A, B)$ be a map of degree 1 of Poincaré pairs of dimension $m=4k$. Suppose $(f|Y)^*: H^*(B; \mathbb{Q})\to H^*(Y; \mathbb{Q})$ is an isomorphism, and that f is cobordant* rel Y *to $f':(X', Y)\to(A, B)$ such that $f'^*: H^*(A; \mathbb{Q})\to H^*(X'; \mathbb{Q})$ is an isomorphism. Then $I(f)=0$.*

Proof. Let U be the cobordism rel Y between X and X' so that $\partial U = X \cup X'$, $X \cap X' = Y$, $(U, \partial U)$ a Poincaré pair of dimension $m+1$, orientations compatible, and $F:(U, Y) \to (A, B)$ such that $F|X = f$, $F|X' = f'$. We may consider F as a map of degree 1,

$$G:(U, X \cup X') \to (A \times I, A \times 0 \cup B \times I \cup A \times 1).$$

By (III.2.4), $I(G|X \cup X') = 0$, and by (III.2.5) $I(G|X \cup X') = I(f) - I(f')$. Now $I(f') = 0$ since f'^* is an isomorphism, and hence $I(f) = 0$. □

§ 3. Normal Maps, Wu Classes, and the Definition of σ for $m = 4l$

Let (X, Y) be a $\mathbb{Z}_2$-Poincaré pair of dimension m. Define a linear map

$$l_i : H^{m-i}(X, Y; \mathbb{Z}_2) \to \mathbb{Z}_2$$

by $l_i(x) = (\mathrm{Sq}^i x)[X]$ where $[X] \in H_m(X, Y; \mathbb{Z}_2)$ is the orientation class. By Poincaré duality, $H^i(X; \mathbb{Z}_2) \otimes H^{m-i}(X, Y; \mathbb{Z}_2) \xrightarrow{\cup} \mathbb{Z}_2$, $(x, y) = (x \cup y)[X]$ is a non-singular pairing, so that $H^i(X, \mathbb{Z}_2)$ is isomorphic, using this pairing to $\mathrm{Hom}(H^{m-i}(X, Y; \mathbb{Z}_2), \mathbb{Z}_2)$ and hence $l_i(x) = (x, v_i)$ for some $v_i \in H^i(X; \mathbb{Z}_2)$, any $x \in H^{m-i}(X, Y; \mathbb{Z}_2)$.

III.3.1 Definition. $V = 1 + v_1 + v_2 + \cdots$ *is the Wu class of* X, $v_i \in H^i(X; \mathbb{Z}_2)$.

III.3.2 Proposition. *Let* (X, Y), (A, B) *be* $\mathbb{Z}_2$*-Poincaré pairs of dimension* m, $f:(X, Y) \to (A, B)$ *a map of degree 1* (mod 2), *so* $f_*[X] = [A]$. *Then* $v_i(X) = \bar{v}_i + f^*(v_i(A))$, *where* $v_i(X) \in H^i(X; \mathbb{Z}_2)$, $v_i(A) \in H^i(A; \mathbb{Z}_2)$ *are the i-th Wu classes and* $\bar{v}_i \in K^i(X) \cong (\mathrm{kernel}\, \alpha^*)^i$, $\alpha^* : H^i(X; \mathbb{Z}_2) \to H^i(A; \mathbb{Z}_2)$ *the natural splitting map for* f^*.

Proof. Let $x \in H^{m-i}(A, B; \mathbb{Z}_2)$. Then $f^*(\mathrm{Sq}^i x) = \mathrm{Sq}^i(f^* x)$, so

$$\begin{aligned}(f^* x, v_i(X)) &= \mathrm{Sq}^i(f^*(x))\,[X] = (f^*(\mathrm{Sq}^i x))\,[X] = (\mathrm{Sq}^i x)\,(f_*[X]) \\ &= (\mathrm{Sq}^i x)\,[A] = (x, v_i(A)).\end{aligned}$$

Since $(x, y) = (f^* x, f^* y)$, we have $(f^* x, v_i(X)) = (f^* x, f^* v_i(A))$, any $x \in H^{m-1}(A, B; \mathbb{Z}_2)$, so $v_i(X) - f^* v_i(A) \in$ annihilator $f^* H^i(A, B; \mathbb{Z}_2) = (\ker \alpha^*)^i = K^i(X; \mathbb{Z}_2)$ by (I.2.9). □

III.3.3 Proposition. *With notation as in* (III.3.2), *suppose that* $m = 2q$. *Then the pairing* $(\,,)$ *on* $K^q(X, Y; \mathbb{Z}_2)$ *is symplectic* ($(x, x) = 0$ *all* x) *if and only if* $f^* v_q(A) = v_q(X)$.

Proof. $(x,x)=x^2[X]=(\mathrm{Sq}^q x)[X]=(x\cup v_q(X))[X]$ for $x\in H^q(X,Y;\mathbb{Z}_2)$, and since $K^q(X,Y;\mathbb{Z}_2)$ and $(\mathrm{image}\, f^*)^q$ are orthogonal by (I.2.9), $(x, f^* v_i(A))=0$ for $x\in K^q(X;\, Y;\, \mathbb{Z}_2)$. Hence for $x\in K^q(X;\, Y;\, \mathbb{Z}_2)$, $(x,\, x) =(x,\, \bar{v}_q)$ by (III.3.2). Then $(x,\, x)=0$ if and only if $\bar{v}_q=v_q(X)-f^*v_q(A) = 0$. □

III.3.4 Corollary. *Let (X, Y), (A, B) be oriented Poincaré duality pairs of dimension $m=4l$, and let $f:(X, Y)\to(A, B)$ be a map of degree 1. If $f^*v_{2l}(A)=v_{2l}(X)$, then the pairing $(x,y)=(x\cup y)[X]$, ($x,y\in K^*(X,Y)/$Torsion, $[X]\in H_m(X, Y)$ the orientation class) is even.*

Proof. By (III.3.3), $f^*v_{2l}(A)=v_{2l}(X)$ implies that the pairing $(\;,\;)_2$, $(x,\, y)_2=(x\cup y)[X]$, $x,\, y\in K^{2l}(X,\, Y;\mathbb{Z}_2)$ is symplectic. If

$$\eta: H^*(X,\, Y)\to H^*(X,\, Y;\mathbb{Z}_2)$$

is induced by reduction mod 2, then

$$(\eta x,\eta y)_2=(\eta x\cup\eta y)[X]=(\eta(x\cup y))[X]=(x,y)\,\mathrm{mod}\,2,\ \text{for } x,y\in K^{2l}(X,Y).$$

Since $(\;,\;)_2$ is symplectic, $(\eta x,\, \eta x)_2=0$, so $(x,\, x)$ is even for $x\in K^{2l}(X,\, Y)$. □

III.3.5 Corollary. *Let (X, Y), (A, B) be oriented Poincaré pairs of dimension $m=4l$, $f:(X, Y)\to(A, B)$ a map of degree 1 such that $(f|Y)_*: H_*(Y)\to H_*(B)$ is an isomorphism. If $f^*(v_{2l}(A))=v_{2l}(X)$, then $I(f)$ is divisible by 8.*

Proof. By (III.3.4), $(\,,)$ is even on $K^{2l}(X, Y)$/Torsion, and $(f|Y)_*$ an isomorphism implies $(\,,)$ is non-singular (see III, § 2). Hence by (III.1.4), signature of $(\,,)$ is divisible by 8. Then

$$K^{2l}(X,\, Y;\mathbb{Q})=(K^{2l}(X,\, Y)/\mathrm{Torsion})\otimes\mathbb{Q},$$

so signature $(\,,)\doteq I(f)$. □

Now we investigate the Wu class V and show that normal maps preserve the Wu class.

Let (X, Y) be a pair, and let ξ^k be a fibre space over X with fibre F such that $H_*(F;\mathbb{Z}_2)=H_*(S^{k-1};\mathbb{Z}_2)$. We recall from (I, § 4) (see also II. § 2) if we set $T(\xi)=X\cup cE(\xi)$ using the projection of ξ as the attaching map, then there is a Thom class $U\in H^k(T(\xi);\mathbb{Z}_2)$ such that

$$\cup U: H^q(X;\mathbb{Z}_2)\to H^{q+k}(T(\xi);\mathbb{Z}_2)$$
$$\cup U: H^q(X,\, Y;\mathbb{Z}_2)\to H^{q+k}(T(\xi),\, T(\xi\,|\,Y);\mathbb{Z}_2)$$
$$\cap U: H_s(T(\xi),\, T(\xi\,|\,Y);\mathbb{Z}_2)\to H_{s-k}(X,\, Y;\mathbb{Z}_2)$$
$$\cap U: H_s(T(\xi);\mathbb{Z}_2)\to H_{s-k}(X;\mathbb{Z}_2)$$

are isomorphisms. Let $h:\pi_r(A, B)\to H_r(A, B;\mathbb{Z}_2)$ be the Hurewicz homomorphism mod 2.

III.3.6 Proposition. *Let (X, Y) be a $\mathbb{Z}_2$-Poincaré pair of dimension m, ξ^k a fibre space over X with fibre F a $\mathbb{Z}_2$ homology $(k-1)$-sphere, $\alpha\in\pi_{m+k}(T(\xi), T(\xi|Y))$ such that $(h(\alpha))\cap U=[X]\in H_m(X, Y;\mathbb{Z}_2)$, the fundamental class of (X, Y). Then $V(X)\cup U=\mathrm{Sq}^{-1}(U)$.*

Proof. $V=V(X)\in H^*(X;\mathbb{Z}_2)$ is characterized by the equation $(x, V)=(\mathrm{Sq}\,x)[X]$, any $x\in H^*(X, Y;\mathbb{Z}_2)$. If γ is a cohomology operation which raises degree, since $h(\alpha)$ is spherical, $(\gamma z)(h(\alpha))=0$, any $z\in H^*(T(\xi), T(\xi|Y);\mathbb{Z}_2)$. Now $\mathrm{Sq}^{-1}=1+\chi(\mathrm{Sq}^1)+\cdots$, so

$$(\mathrm{Sq}^{-1}z)(h(\alpha))=(z)(h(\alpha)).$$

Hence

$$(\mathrm{Sq}\,x)[X]=(\mathrm{Sq}\,x)(h(\alpha)\cap U)=(\mathrm{Sq}\,x\cup U)(h(\alpha))=$$
$$(\mathrm{Sq}^{-1}(\mathrm{Sq}\,x\cup U))(h(\alpha))=(x\cup\mathrm{Sq}^{-1}U)(h(\alpha))=(x\cup V')[X]=(x, V')$$

where $\mathrm{Sq}^{-1}U=V'\cup U$. But V is characterized by this equation so $V'=V$ and $\mathrm{Sq}^{-1}U=V\cup U$. $\square$

We recall that the Thom class $U\in H^k(T(\xi);\mathbb{Z}_2)$ is characterized by the fact that $j^*(U)$ generates $H^k(\Sigma F;\mathbb{Z}_2)=\mathbb{Z}_2$, where $j:\Sigma F\to T(\xi)$ is the inclusion of the Thom complex over a point into the whole Thom complex.

III.3.7 Proposition. *Let $b:\xi\to\xi'$ be a map of fibre spaces over $f:X\to X'$, where ξ, ξ' have fibre F, $H_*(F;\mathbb{Z}_2)=H_*(S^{k-1};\mathbb{Z}_2)$. Then b induces a map of Thom complexes $T(b): T(\xi)\to T(\xi')$, and $T(b)^*(U')=U$, $U'=$ Thom class in $H^k(T(\xi');\mathbb{Z}_2)$, U the Thom class in $H^k(T(\xi);\mathbb{Z}_2)$.*

Proof. Let E, E' be the total spaces of ξ, ξ' respectively, so that the following diagram commutes:

$$\begin{array}{ccccc} F & \longrightarrow & E & \xrightarrow{\pi} & X \\ \downarrow{\scriptstyle 1} & & \downarrow{\scriptstyle b} & & \downarrow{\scriptstyle f} \\ F & \longrightarrow & E' & \xrightarrow{\pi'} & X'. \end{array}$$

Hence f, b induce

$$T(b): X\bigcup_{\pi} cE\to X'\bigcup_{\pi'} cE'$$

and the diagram

$$\begin{array}{ccc} \Sigma F & \xrightarrow{j} & T(\xi) \\ \downarrow{\scriptstyle 1} & & \downarrow{\scriptstyle T(b)} \\ \Sigma F & \xrightarrow{j'} & T(\xi') \end{array}$$

commutes. Hence $j^*T(b)^*(U')=j'^*(U')$ so that $j^*T(b)^*(U')$ generates $H^k(\Sigma F;\mathbb{Z}_2)$ and hence $T(b)^*(U')=U$. $\square$

III.3.8 Corollary. *Let $(X, Y), (A, B)$ be $\mathbb{Z}_2$-Poincaré pairs of dimension m, ξ' a fibre space over A with fibre F a $(k-1)$ dimensional $\mathbb{Z}_2$-homology sphere. Let $f:(X, Y)\to(A, B)$ be a map of degree* 1 mod 2, *and let $\xi=f^*(\xi')$. Suppose there is an element $\alpha\in\pi_{m+k}(T(\xi), T(\xi\,|\,Y))$ such that $h(\alpha)\cap U=[X]$, the fundamental class in $H_m(X, Y;\mathbb{Z}_2)$, $U\in H^k(T(\xi);\mathbb{Z}_2)$ the Thom class, h the Hurewicz homomorphism. Then $f^*(V(A))=V(X)$, in particular $f^*v_q(A)=v_q(X)$, all q.*

Proof. By (III.3.7), if $b:\xi\to\xi'$ is the natural map, $T(b)^*U'=U$. Setting $V(X)=V$, $V(A)=V'$, we have, using (III.3.6), $T(b)^*(V'\cup U') =f^*V'\cup T(b)^*U'=f^*(V')\cup U=T(b)^*(\mathrm{Sq}^{-1}U')=\mathrm{Sq}^{-1}T(b)^*U'=\mathrm{Sq}^{-1}U =V\cup U$. Hence $f^*V'=V$. □

III.3.9 Theorem. *Let $(X, Y), (A, B)$ be oriented Poincaré pairs of dimension $m=4l$, let $f:(X, Y)\to(A, B)$ be a map of degree 1 such that $(f\,|\,Y)_*$ is an isomorphism, and let ξ' be a fibre space over A with fibre F a $\mathbb{Z}_2$-homology $(k-1)$-sphere. Set $\xi=f^*\xi'$ and suppose there is $\alpha\in\pi_{m+k}(T(\xi), T(\xi\,|\,Y))$ such that $h(\alpha)\cap U=$ orientation class of (X, Y) mod 2, (where $U\in H^k(T(\xi);\mathbb{Z}_2)$ is the Thom class, $h=$ Hurewicz homomorphism). Then $I(f)$ is divisible by* 8.

Proof. By (III.3.8), $f^*v_{2l}(A)=v_{2l}(X)$, so by (III.3.5), $I(f)$ is divisible by 8. □

Let (f, b) be a normal map, $f:(M, \partial M)\to(A, B)$ a map of degree 1, M^m a smooth oriented m-manifold with boundary, (A, B) an oriented Poincaré pair of dimension m, $m=4l$, and $b:v\to\eta$ is a linear bundle map covering f, v the normal bundle of $(M, \partial M)\subset(D^{m+k}, S^{m+k-1})$, η a k-plane bundle over A.

III.3.10 Coroïlary. *If (f, b) is a normal map with $(f\,|\,\partial M)_*$ an isomorphism, then $I(f)$ is divisible by* 8.

Proof. The pair (f, b) satisfies the conditions of (III.3.9) where $\xi'=\eta$ is a linear bundle over (A, B). □

III.3.11 Definition. Let (f, b) be a normal map $f:(M, \partial M)\to(A, B)$, etc. with $(f\,|\,\partial M)_*$ an isomorphism, $m=$ dimension $M=4l$. Define $\sigma(f, b)=\frac{1}{8}I(f)\in\mathbb{Z}$.

Then (II.1.1), the Invariant Theorem, follows for $m=4l$ from (III.2.9). The Addition Property (II.1.4) follows from (III.2.5), the Cobordism Property (II.1.5) follows from (III.2.4). For the Index Property (II.1.6) we note that by (III.2.8), $8\sigma(f, b)=I(f)=\mathrm{Index}\,M-\mathrm{Index}\,A$, and by the Hirzebruch Index Theorem [30],

$$\begin{aligned}\mathrm{Index}\,M&=\big(L_l(p_1(\tau_M), \dots, p_l(\tau_M))\big)([M]) =\big(L_l(p_1(v^{-1}), \dots, p_l(v^{-1}))\big)[M]\\ &=\big(L_l(p_1(\eta^{-1}), \dots, p_l(\eta^{-1}))\big)[A]\end{aligned}$$

where $[A]$ is the orientation class in $H_{4l}(A, B)$. This proves the Index Property (II.1.6) (recalling that $A=X, \eta=\xi, k=l$ to retrieve the original notation).

§ 4. The Invariant $c(f, b)$

Let (X, Y) and (A, B) be oriented Poincaré pairs of dimension $m=2q$, and let $f:(X, Y)\to(A, B)$ be a map of degree 1. Let ξ be the Spivak normal fibre space of (X, Y) and η that of (A, B), and let $\alpha\in\pi_{m+k}(T(\xi), T(\xi\,|\,Y))$, $\beta\in\pi_{m+k}(T(\eta), T(\eta\,|\,B))$ be the elements defined in (I.4.4) such that $h(\alpha)\cap U_\xi=[X], h(\beta)\cap U_\eta=[A]$, where $U_\xi\in H^k(T(\xi))$, $U_\eta\in H^k(T(\eta))$ are the respective Thom classes, h the Hurewicz homomorphism. Let $b:\xi\to\eta$ be a map of fibre spaces over f. We shall call the pair (f, b) a normal map of Poincaré pairs, (compare II, § 1).

Normal cobordism and normal cobordism rel B is defined analogously (c.f. II, § 1).

By (I.4.15), $T(\xi)$ is $(m+k+s)$-dual to $T(\varepsilon^s)/T(\varepsilon^s\,|\,Y)\cong\Sigma^s(X/Y)$, ($\varepsilon^s$ = trivial bundle over X) and $T(\eta)$ is $(m+k+s)$-dual to $\Sigma^s(A/B)$. Hence for a normal map (f, b), $T(b): T(\xi)\to T(\eta)$ is $(m+k+s)$-dual (k very large) to a map $g:\Sigma^s(A/B)\to\Sigma^s(X/Y)$.

In fact we will only use mod 2 properties of these things in defining $c(f, b)$. Thus it is possible to weaken the hypotheses, for example to (X, Y) and (A, B) $\mathbb{Z}_2$-Poincaré pairs, with appropriate fibre spaces with $\mathbb{Z}_2$ homology spheres as fibre in place of the Spivak normal fibre space etc.

The map g may be related to certain maps constructed in Chapter I. Recall that in Chapter I, § 2 we defined $\alpha^*: H^*(X/Y)\to H^*(A/B)$ by $[A]\cap\alpha^*(x)=f_*([X]\cap x)$, for $x\in H^*(X/Y)$, $[X], [A]$ the orientation classes of $(X, Y), (A, B)$ respectively. Let $\Sigma^*: H^q(K)\to H^{q+s}(\Sigma^s K)$ be the suspension isomorphism for any space K.

III.4.1 Theorem. $g^*\Sigma^*=\Sigma^*\alpha^*$.

Proof. By (I.4.14) the condition that $T(b)$ and g are $(m+k+s)$ dual is equivalent to the commutativity (up to homotopy) of the following diagram:

$$\begin{array}{ccc} S^{m+k+s} & \xrightarrow{\gamma} & T(\xi)\wedge\Sigma^s(X/Y) \\ \downarrow{\scriptstyle\gamma'} & & \downarrow{\scriptstyle T(b)\wedge 1} \\ T(\eta)\wedge\Sigma^s(A/B) & \xrightarrow{1\wedge g} & T(\eta)\wedge\Sigma^s(X/Y). \end{array}$$

Here γ and γ' are defined as in the proof of (I.4.15). In particular $\gamma_*(\iota)\cap U_\xi\cap U=\Delta_*[X]$, and $\gamma'_*(\iota)\cap U_\eta\cap U'=\Delta'_*[A]$, where $\iota\in H_{m+k+s}(S^{m+k+s})$ is the generator, $U\in H^s(\Sigma^s X_+)$, $U'\in H^s(\Sigma^s A_+)$ are Thom classes, (considering $\Sigma^s(X/Y)$ as $T(\varepsilon)/T(\varepsilon|Y)$ etc.) so that $\Sigma^* x = x \cup U$, $x \in H^*(X/Y)$, etc. and $\Delta: (X, Y) \to X \times (X, Y)$ and $\Delta': (A, B) \to A \times (A, B)$ are the diagonal maps (see diagram (*) in the proof of (I.4.15)).

It follows that

$$\gamma'_*(\iota)/g^*(x \cup U) = T(b)_*(\gamma_*(\iota)/(x \cup U)).$$

But $(\gamma_*(\iota)/(x \cup U)) \cap U_\xi = (\gamma_*(\iota) \cap U_\xi \cap U)/x = \Delta_*[X]/x = [X] \cap x$. But $U_\xi = T(b)^* U_\eta$ so that

$$(T(b)_*(\gamma_*(\iota)/(x \cup U))) \cap U_\eta = f_*((\gamma_*(\iota)/x \cup U) \cap U_\xi) = f_*([X] \cap x).$$

Now in a similar way, if $g^*(x \cup U) = y \cup U'$, it follows that

$$(\gamma'_*(\iota)/g^*(x \cup U)) \cap U_\eta = (\gamma'_*(\iota)/(y \cup U')) \cap U_\eta$$
$$= \gamma'_*(\iota) \cap U_\eta \cap U'/y = \Delta'_*[A]/y = [A] \cap y.$$

Hence $[A] \cap (\Sigma^{*-1} g^* \Sigma^* x) = f_*([X] \cap x)$, so $\Sigma^{*-1} g^* \Sigma^* x = \alpha^* x$, so $g^* \Sigma^* = \Sigma^* \alpha^*$. □

III.4.2 Corollary. *If (f, b) is a normal map of Poincaré pairs, $f: (X, Y) \to (A, B)$, then $\Sigma^s f: \Sigma^s(X/Y) \to \Sigma^s(A/B)$ is a domination for sufficiently large s, i.e. $\Sigma^s f$ has a homotopy right inverse.*

Proof. Consider $(\Sigma^s f) g: \Sigma^s(A/B) \to \Sigma^s(A/B)$. Then

$$\Sigma^{*-1}((\Sigma^s f) g)^* \Sigma^* = \Sigma^{*-1} g^* (\Sigma^s f)^* \Sigma^* = \Sigma^{*-1} g^* \Sigma^* f^*$$
$$= \Sigma^{*-1} \Sigma^* \alpha^* f^* = \alpha^* f^* = 1$$

by (III.4.1) and (I.2.5). Hence $h = (\Sigma^s f) g$ induces isomorphism on $H^*(\Sigma^s(A/B))$ and hence on $H_*(\Sigma^s(A/B))$. It follows that h is a homotopy equivalence and therefore $(g h^{-1})$ is a homotopy right inverse for $(\Sigma^s f)$. □

III.4.3 Corollary. *For normal maps the splitting map $\alpha^*: H^*(X/Y) \to H^*(A/B)$ commutes with stable cohomology operations. In particular $\alpha^* \mathrm{Sq}^i = \mathrm{Sq}^i \alpha^*$.*

Proof. $\Sigma^* \alpha^* = g^* \Sigma^*$, so $\alpha^* = \Sigma^{*-1} g^* \Sigma^*$, and since g^* and Σ^* commute with stable cohomology operations, so does α^*. □

This gives us another proof of the fact that $(x, x) = 0$ for $x \in K^q(X, Y; \mathbb{Z}_2)$ (see III. § 3). For $K^q(X, Y) = (\ker \alpha^*)^q$, and

$$(x, x) = x^2[X] = (\mathrm{Sq}^q x)[X] = (\mathrm{Sq}^q x)(\alpha_*[A])$$
$$= (\alpha^* \mathrm{Sq}^q x)[A] = (\mathrm{Sq}^q \alpha^* x)[A] = 0.$$

Now we shall use the map g and (III.4.1) to construct a quadratic form on $K^q(X, Y; \mathbb{Z}_2)$. The construction follows that in [7].

Recall that (see [55]) the Eilenberg-Mac Lane space $K(\mathbb{Z}_2, q)$ is a space such that $\pi_i(K(\mathbb{Z}_2, q)) = 0$ for $i \neq q$ and $\pi_q(K(\mathbb{Z}_2, q)) = \mathbb{Z}_2$. It is a simple consequence of obstruction theory that this condition defines the homotopy type of $K(\mathbb{Z}_2, q)$ uniquely in the category of CW complexes and that homotopy classes of maps of a CW-complex L into $K(\mathbb{Z}_2, q)$ are in one-to-one correspondence with elements $x \in H^q(L; \mathbb{Z}_2)$, i.e., $e: [L, K(\mathbb{Z}_2, q)] \to H^q(L; \mathbb{Z}_2)$ is a $1-1$ correspondence where $e(f) = f^*(\iota)$, ι is the generator of $H^q(K(\mathbb{Z}_2, q); \mathbb{Z}_2) = \mathbb{Z}_2$.

Let $x \in H^q(X/Y; \mathbb{Z}_2)$, $x \in (\text{kernel}\, \alpha^*)$ so that $g^*(\Sigma^* x) = 0$, and let $\varphi: X/Y \to K(\mathbb{Z}_2, q)$ be a map such that $\varphi^*(\iota) = x$. Take $h = (\Sigma^s \varphi) g$

$$\Sigma^s(A/B) \xrightarrow{\ g\ } \Sigma^s(X/Y) \xrightarrow{\Sigma^s \varphi} \Sigma^s K(\mathbb{Z}_2, q).$$

Now we recall the definition of functional cohomology operation, due to Steenrod [61]. Let $\omega: H^n(X; G) \to H^{n+k}(X; G')$ be a stable cohomology operation, (e.g. Sq^k) and let $f: K \to L$ be a map of spaces. Let $x \in H^n(L; G)$ such that

(i) $f^*(x) = 0$ and

(ii) $\omega(x) = 0$.

Then the functional operation $\omega_f(x)$ is defined as an element of

$$\frac{H^{n+k-1}(K; G')}{\omega H^{n-1}(K; G) + f^* H^{n+k-1}(L; G')}$$

defined using the exact sequence of f:

$$\begin{array}{ccccccccc}
 & \longrightarrow H^{n-1}(K;G) & \xrightarrow{\ \delta\ } & H^n(f;G) & \xrightarrow{j^*} & H^n(L;G) & \xrightarrow{f^*} & H^n(K;G) \longrightarrow \\
 & \downarrow \omega & & \downarrow \omega & & \downarrow \omega & & \downarrow \omega \\
H^{n+k-1}(L;G') \xrightarrow{f^*} & H^{n+k-1}(K;G') & \xrightarrow{\delta} & H^{n+k}(f;G') & \xrightarrow{j^*} & H^{n+k}(L;G') & \xrightarrow{f^*} & H^{n+k}(K;G').
\end{array}$$

Since $f^* x = 0$, by exactness $x = j^* y$, $y \in H^n(f; G)$. Now

$$j^* \omega y = \omega j^* y = \omega x = 0,$$

so by exactness, $\omega y = \delta z$, $z \in H^{n+k-1}(K; G')$. Then z represents $\omega_f(x)$. Now $\delta f^* = 0$, so at the last step z is only well defined $\mathrm{mod}\, f^* H^{n+k-1}(L; G')$. Also, $j^* \delta = 0$, so y is only well defined $\mathrm{mod}\, \delta H^{n-1}(K; G)$, so that ωy is only well defined $\mathrm{mod}\, \omega \delta H^{n-1}(K; G) = \delta \omega H^{n-1}(K; G)$, (since ω is a stable cohomology operation). Hence z is only well defined

$$\mathrm{mod}\, f^* H^{n+k-1}(L; G') + \omega H^{n-1}(K; G).$$

Returning to our situation, we have

$$\Sigma^s(A/B) \xrightarrow{\ g\ } \Sigma^s(X/Y) \xrightarrow{\Sigma^s \varphi} \Sigma^s K(\mathbb{Z}_2, q),\ h = (\Sigma^s \varphi) g,\ \varphi^* \iota = x \in H^q(X/Y; \mathbb{Z}_2),$$

where $h^*(\Sigma^s \iota) = 0$. Then the operation $\mathrm{Sq}^{q+1}(\iota) = 0$ in $H^*(K(\mathbb{Z}_2, q); \mathbb{Z}_2)$ since $\dim \iota = q$ and $\mathrm{Sq}^k(c) = 0$ if $\dim c < k$. Hence we may define the functional operation $\mathrm{Sq}_h^{q+1}(\Sigma^s(\iota)) \in H^{2q+s}(\Sigma^s(A/B); \mathbb{Z}_2)$

$$\mathrm{mod}\, h^*(H^{2q+s}(\Sigma^s K(\mathbb{Z}_2, q); \mathbb{Z}_2)) + \mathrm{Sq}^{q+1} H^{q+s-1}(\Sigma^s(A/B); \mathbb{Z}_2).$$

III.4.4 Lemma. *The indeterminacy* image h^* + image $\mathrm{Sq}^{q+1} = 0$. (Compare [7, (1.1)].)

Proof. Since $H^{q+s-1}(\Sigma^s(A/B);\mathbb{Z}_2) = \Sigma^s H^{q-1}(A/B;\mathbb{Z}_2)$, and since Sq^{q+1} is identically zero on $H^{q-1}(A/B;\mathbb{Z}_2)$ for dimensional reasons, it follows that image $\mathrm{Sq}^{q+1} = 0$ in $H^{2q+s}(\Sigma^s(A/B);\mathbb{Z}_2)$. By a theorem of Serre [52], $H^{2q}(K(\mathbb{Z}_2, q);\mathbb{Z}_2)$ is obtained by acting with the Steenrod algebra $\mathscr{A}_2$ on $\iota \in H^q(K(\mathbb{Z}_2, q);\mathbb{Z}_2)$. It follows that $H^{2q+s}(\Sigma^s K(\mathbb{Z}_2, q);\mathbb{Z}_2)$ is obtained from $\Sigma^s(\iota)$ by action of $\mathscr{A}_2$. Then $h^*(\Sigma^s(\iota)) = 0$ implies

$$h^*(a\Sigma^s(\iota)) = ah^*\Sigma^s(\iota) = 0$$

for $a \in \mathscr{A}_2$, and hence $h^* H^{2q+s}(\Sigma^s K(\mathbb{Z}_2, q);\mathbb{Z}_2) = 0$. □

III.4.5 Definition. $\psi: K^q(X, Y;\mathbb{Z}_2) \to \mathbb{Z}_2$ by $\psi(x) = (\mathrm{Sq}_h^{q+1}(\Sigma^s(\iota)))(\Sigma^s[A])$ where h etc., is as above.

III.4.6 Proposition. *ψ is a quadratic form on $K^q(X, Y;\mathbb{Z}_2)$ and its associated bilinear form is $(\ ,\ )$, where $(x, y) = (x \cup y)[X]$, for $x, y \in K^q(X, Y;\mathbb{Z}_2)$.*

Proof. We outline the proof briefly referring to [7, (1.4)] for the details. Set $M = X/Y$.

Let $x_1, x_2 \in H^q(M;\mathbb{Z}_2)$ such that $\alpha^* x_1 = \alpha^* x_2 = 0$. Let $\varphi_i: M \to K(\mathbb{Z}_2, q)$ be such that $\varphi_i^*(\iota) = x_i$, $i = 1, 2$, and let φ be the composite

$$M \xrightarrow{\Delta} M \times M \xrightarrow{\varphi_1 \times \varphi_2} K \times K \xrightarrow{\mu} K,$$

$K = K(\mathbb{Z}_2, q)$, $\Delta(m) = (m, m)$, $m \in M$, μ is the multiplication map in K. Then $\varphi^*(\iota) = x_1 + x_2$.

Then $\Sigma^s\varphi : \Sigma^s M \to \Sigma^s K$ is the composite of the suspended maps. For any X and Y, we have natural homotopy equivalences

$$\varrho : \Sigma X \vee \Sigma Y \vee \Sigma(X \wedge Y) \to \Sigma(X \times Y)$$

where $\varrho = \Sigma i + \Sigma j + h(1)$, $i: X \to X \times Y$ by $i(x) = (x, *)$, $j: Y \to X \times Y$, $j(y) = (*, y)$, $*$ denoting base point, and $h(1)$ is the Hopf construction on the identity $1: X \times Y \to X \times Y$, (see [55] and [59]). Here $+$ denotes the sum of maps in the group of homotopy classes of

$$[\Sigma X \vee \Sigma Y \vee \Sigma(X \wedge Y), \Sigma(X \times Y)].$$

It then follows from naturality that $\Sigma^s A/B \xrightarrow{\eta} \Sigma^s M \xrightarrow{\Sigma^s\varphi} \Sigma^s K$ is the sum of three maps $\xi_1 + \xi_2 + \gamma$ where $\xi_i = \Sigma^s\varphi_i \circ \eta$, and

$$\gamma = \Sigma^{s-1} h(\mu) \circ \Sigma^s(\varphi_1 \wedge \varphi_2) \circ \Sigma^s\bar{\Delta} \circ \eta,$$

where $\bar{\Delta}$ is the composite

$$M \xrightarrow{\Delta} M \times M \to M \wedge M, \qquad f_1 \wedge f_2 : M \wedge M \to K \wedge K,$$

$h(\mu)$ is the Hopf construction on $\mu: K \times K \to K$.

It is an easy exercise to show that

$$\mathrm{Sq}^{q+1}_{(\xi_1+\xi_2+\gamma)}(\Sigma^s\iota)=\mathrm{Sq}^{q+1}_{\xi_1}(\Sigma^s\iota)+\mathrm{Sq}^{q+1}_{\xi_2}(\Sigma^s\iota)+\mathrm{Sq}^{q+1}_{\gamma}(\Sigma^s\iota).$$

Setting $\zeta=\Sigma^{s-1}h(\mu):\Sigma^s K\wedge K\to\Sigma^s K$, we have $\zeta^*(\Sigma^s\iota)=0$, so $\mathrm{Sq}^{q+1}_{\zeta}(\Sigma^s(\iota))$ is defined, and since the indeterminancy of $\mathrm{Sq}^{q+1}_{\zeta}$ is zero in $H^{2q+s}(\Sigma^s K\wedge K;\mathbb{Z}_2)$ (since $H^i(\Sigma^s K\wedge K;\mathbb{Z}_2)$ is zero for $i<2q+s$) it follows that

$$\mathrm{Sq}^{q+1}_{\gamma}(\Sigma^s(\iota))=\eta^*\circ(\Sigma^s\bar{\Delta})^*\circ(\Sigma^s\varphi_1\wedge\varphi_2)^*\,\mathrm{Sq}^{q+1}_{\zeta}(\Sigma^s(\iota)).$$

Now $((\Sigma^s\varphi_1\wedge\varphi_2)\cdot(\Sigma^s\bar{\Delta}))^*=(\Sigma^s((\varphi_1\wedge\varphi_2)\circ\bar{\Delta}))^*$ and $((\varphi_1\wedge\varphi_2)\circ\bar{\Delta})^*(\iota\wedge\iota)$ $=x_1\cup x_2\in H^{2q}(X/Y;\mathbb{Z}_2)$ (recalling we have set $M=X/Y$). Now $\mathrm{Sq}^{q+1}_{h(\mu)}(\Sigma(\iota))=\Sigma(\iota\wedge\iota)\in H^{2q+1}(\Sigma K\wedge K;\mathbb{Z}_2)$, as is easily shown by an argument analogous to [59, (5.3)]. It follows now that

$$\psi(x_1+x_2)=\psi(x_1)+\psi(x_2)+(x_1\cup x_2)[X].\quad\square$$

Now if $(f|Y)^*:H^*(B;\mathbb{Z}_2)\to H^*(Y;\mathbb{Z}_2)$ is an isomorphism, it follows from (I.2.9) that (,) is non-singular on $K^q(X,Y;\mathbb{Z}_2)$ $(\cong K^q(X;\mathbb{Z}_2))$. Then, by III. § 1, the Arf invariant $c(\psi)$ is defined.

III.4.7 Definition. Let (f,b) be a normal map of Poincaré complexes $f:(X,Y)\to(A,B)$, and suppose that $(f|Y)^*:H^*(B;\mathbb{Z}_2)\to H^*(Y;\mathbb{Z}_2)$ is an isomorphism. Then define the *Kervaire invariant* $c(f,b)=c(\psi)$, the Arf invariant of $\mathbb{Z}$.

Now we will proceed to develop the properties of $c(f,b)$.

Let (f,b) be a normal map, $f:(X,Y)\to(A,B)$ etc., and suppose in addition that Y and B are sums of Poincaré pairs along the boundaries and f sends summands into summands. In particular, we suppose $Y=Y_1\cup Y_2$, $Y_0=Y_1\cap Y_2$, $B=B_1\cup B_2$, $B_0=B_1\cap B_2$, $f(Y_i)\subset B_i$, and that (B_i,B_0), (Y_i,Y_0) are Poincaré pairs compatibly oriented with (X,Y) and (A,B) (see I. § 3). If ξ,η are the Spivak normal fibre spaces of (X,Y) and (A,B) then $\xi|Y_i$, $\eta|B_i$ are the corresponding Spivak normal fibre spaces, so that if $f_i=f|Y_i$, $b_i=b|(\xi|Y_i)$ then (f_i,b_i) are all normal maps, $i=0,1,2$.

We note that if $f_2^*:H^*(B_2;\mathbb{Z}_2)\to H^*(Y_2;\mathbb{Z}_2)$ is an isomorphism then it follows from (I.2.6) and (I.2.7) that $f_0^*:H^*(B_0;\mathbb{Z}_2)\to H^*(Y_0;\mathbb{Z}_2)$ is an isomorphism.

III.4.8 Theorem. *Let (f,b) be a normal map as above, so that $f|Y$ is the sum of f_1 and f_2 on Y_1 and Y_2 etc. Suppose $f_2^*:H^*(B_2;\mathbb{Z}_2)\to H^*(Y_2;\mathbb{Z}_2)$ is an isomorphism. Then $c(f_1,b_1)=0$.*

This theorem has the following corollaries:

III.4.9 Corollary. *If (f,b) is a normal map and is normally cobordant* rel Y *to (f',b'), $f'^*:H^*(A,B;\mathbb{Z}_2)\to H^*(X',Y;\mathbb{Z}_2)$ an isomorphism, then $c(f,b)=0$.*

III.4.10 Corollary. *If (f, b) is a normal map, $f:(X, Y)\to(A, B)$, then $c(f \mid Y, b \mid (\xi \mid Y))=0$.*

It is clear how to derive the two corollaries, in the first using the normal cobordism as a normal map, and in the second taking $Y_2=\emptyset$.

We will utilize some lemmas.

III.4.11 Proposition. *Let (f, b) be a normal map, $f:(X, Y)\to(A, B)$, and let $g:\Sigma^s(A/B)\to\Sigma^s(X/Y)$ and $g':\Sigma^{s+1}B\to\Sigma^{s+1}Y$ be the duals of $T(b)$ and $T(b \mid (\xi \mid Y))$. Then the diagram below commutes up to homotopy:*

$$\begin{array}{ccc} \Sigma^s(A/B) & \xrightarrow{g} & \Sigma^s(X/Y) \\ \Big\downarrow{\scriptstyle \Sigma^s d'} & & \Big\downarrow{\scriptstyle \Sigma^s d} \\ \Sigma^{s+1} B_+ & \xrightarrow{g'} & \Sigma^{s+1} Y_+ \end{array}$$

where $d: X/Y\to\Sigma Y$, $d': A/B\to\Sigma B$ are the natural maps, (considering X/Y as $X\cup cY$, and smashing X to a point).

Proof. We have a commutative diagram

$$\begin{array}{ccc} T(\xi \mid Y) & \xrightarrow{T(b \mid (\xi \mid Y))} & T(\eta \mid B) \\ \Big\downarrow{\scriptstyle j} & & \Big\downarrow{\scriptstyle j'} \\ T(\xi) & \xrightarrow{T(b)} & T(\eta) \end{array}$$

where j, j' are inclusions. Then the dual diagram commutes:

$$\begin{array}{ccc} \Sigma^s(A/B) & \xrightarrow{g} & \Sigma^s(X/Y) \\ \Big\downarrow{\scriptstyle a'} & & \Big\downarrow{\scriptstyle a} \\ \Sigma^{s+1} \mathrm{B}_+ & \xrightarrow{g'} & \Sigma^{s+1} \mathrm{Y}_+ . \end{array}$$

It remains to show that a is homotopic to $\Sigma^s d$ and a' is homotopic to $\Sigma^s d'$, i.e. that $\Sigma^s d$ and j are dual in S-theory, and similarly for $\Sigma^s d'$ and j'. Then (III.4.11) follows from:

III.4.12 Lemma. *Let (X, Y) be a Poincaré pair, ξ its Spivak normal fibre space. Then the inclusion $j: T(\xi \mid Y)\to T(\xi)$ is $(m+k)$ dual to d, where d is the natural map $d: X/Y\to\Sigma Y$.*

Proof. By (I.4.14) the statement is equivalent to the commutativity up to homotopy of the diagram

$$\begin{array}{ccc} S^{m+k} & \xrightarrow{\varrho} & T(\xi)\wedge(X/Y) \\ \Big\downarrow{\scriptstyle \varrho'} & & \Big\downarrow{\scriptstyle 1\wedge d} \\ T(\xi \mid Y)\wedge\Sigma(Y_+) & \xrightarrow{j\wedge 1} & T(\xi)\wedge\Sigma(Y_+) . \end{array}$$

We recall the definition of ϱ and ϱ' (I.4.15).

Consider the map $b:\xi\to\xi\times\varepsilon^0$, where ε^k is the trivial fibre space of dimension k, covering the diagonal $\Delta:(X,Y)\to X\times(X,Y)$. We have a commutative diagram

$$\begin{array}{ccccc}
S^{m+k-1} & \xrightarrow{\alpha'} & T(\xi\,|\,Y) & \xrightarrow{\omega'} & T(\xi\,|\,Y)\wedge(Y_+) \\
\Big\downarrow & & \Big\downarrow & & \Big\downarrow{\scriptstyle j\wedge 1} \\
 & & T(\xi\,|\,Y) & \xrightarrow{\omega''} & T(\xi)\wedge(Y_+) \\
\Big\downarrow & & \Big\downarrow & & \Big\downarrow{\scriptstyle 1\wedge i} \\
D^{m+k} & \xrightarrow{\alpha} & T(\xi) & \xrightarrow{\omega} & T(\xi)\wedge(X_+)
\end{array}$$

where $T(\xi)\wedge X_+=T(\xi\times\varepsilon^0)$ over $X\times X$, $T(\xi)\wedge(Y_+)=T(\xi\times\varepsilon^0\,|\,X\times Y)$, $T(\xi\,|\,Y)\wedge(Y_+)=T(\xi\times\varepsilon^0\,|\,X\times Y)$, α, α' are the collapsing maps, and ω,ω',ω'' are induced by b and its restrictions.

Then $\gamma'=\omega'\alpha':S^{m+k-1}\to T(\xi\,|\,Y)\wedge(Y_+)$ is the duality map for Y_+ and $T(\xi\,|\,Y)$ while the map of pairs

$$\gamma=(\omega\alpha,\omega''\alpha'):(D^{m+k},S^{m+k-1})\to(T(\xi)\wedge(X_+),\,T(\xi)\wedge(Y_+))$$

represents the duality map for $T(\xi)$ and X/Y when the subspaces are pinched to a point. Now $\omega''\alpha'=(j\wedge 1)\omega'\alpha'$, so we have

$$\partial\{\gamma\}=(j\wedge 1)_*\{\gamma'\},\ \partial:\pi_{m+k}(T(\xi)\wedge(X_+),\,T(\xi)\wedge(Y_+))\to\pi_{m+k-1}(T(\xi)\wedge(Y_+))\,.$$

As $\varrho:D^{m+k}/S^{m+k-1}\to T(\xi)\wedge(X_+)/T(\xi)\wedge(Y_+)=T(\xi)\wedge(X/Y)$ is induced by γ, then $\Sigma\partial\{\gamma\}=\bar{d}_*\{\varrho\}$ from the general properties of homotopy groups, where $\bar{d}:T(\xi)\wedge(X_+)/T(\xi)\wedge(Y_+)\to\Sigma(T(\xi)\wedge(Y_+))$. Rearranging the suspension parameters to make the homotopy equivalence $h:\Sigma(T(\xi)\wedge(Y_+))\to T(\xi)\wedge\Sigma(Y_+)$ then shows that $h\bar{d}=1\wedge d$ and hence

$$(1\wedge d)_*\{\varrho\}=h_*\bar{d}_*\{\varrho\}=h_*\Sigma\partial\{\gamma\}=h_*\Sigma(j\wedge 1)_*\{\gamma'\}=(j\wedge 1)_*\{\varrho'\}\,,$$

since $\Sigma\gamma'=\varrho'$. Hence the diagram commutes. □

The proof of (III.4.8) is based on the following lemma.

III.4.13 Lemma. *Let (f,b) be a normal map, $f:(X,Y)\to(A,B)$, Y of* dim $2q$, *and let $x\in K^q(X;\mathbb{Z}_2)$. Then $\psi(i^*x)=0$, where $i:Y\to X$ is inclusion.*

Proof. If $\varphi':X\to K(\mathbb{Z}_2,q),\varphi'^*(\iota)=x$, then $\psi(i^*x)$ is defined using the composite

$$h:\Sigma^sB\xrightarrow{g'}\Sigma^sY\xrightarrow{\Sigma^si}\Sigma^sX\xrightarrow{\Sigma^s\varphi'}\Sigma^sK(\mathbb{Z}_2\,q)\,.$$

Now $\Sigma^{s-1}d':\Sigma^{s-1}(A/B)\to\Sigma^sB$ is of degree 1, so that

$$\psi(i^*x)=(\mathrm{Sq}_h^{q+1}(\Sigma^s(\iota)))(\Sigma^{s-1}d'_*[A])=(\mathrm{Sq}_{h'}^{q+1}(\Sigma^s(\iota)))\,[A]$$

where $h' = h(\Sigma^{s-1} d')$. Then we have a commutative diagram from (III.4.11)

$$\begin{array}{ccccccc} \Sigma^{s-1}(A/B) & \xrightarrow{g} & \Sigma^{s-1}(X/Y) & & & & \\ \downarrow{\scriptstyle \Sigma^{s-1}d'} & & \downarrow{\scriptstyle \Sigma^{s-1}d} & & & & \\ \Sigma^s B & \xrightarrow{g'} & \Sigma^s Y & \xrightarrow{\Sigma^s i} & \Sigma^s X & \xrightarrow{\Sigma^s \varphi'} & \Sigma^s K(\mathbb{Z}_2, q) \end{array}$$

so that $h' = (\Sigma^s \varphi')(\Sigma^s i) g'(\Sigma^{s-1} d') = (\Sigma^s \varphi')(\Sigma^s i)(\Sigma^{s-1} d) g$. But

$$(\Sigma^s i)(\Sigma^{s-1} d) = \Sigma^{s-1}((\Sigma i) d)$$

and $(\Sigma i) d$ is homotopic to a constant as is clear from the representation

$$\begin{array}{ccccc} X \cup cY & \xrightarrow{d} & cX \cup cY & \xrightarrow{\Sigma i} & cX \cup cY / X \cup cY \\ & & \wr\| & & \wr\| \\ & & \Sigma Y & & \Sigma X \end{array}$$

(i.e. it is the composition of two consecutive terms in the sequence

$$Y \to X \to X/Y \to \Sigma Y \to \Sigma X \to \cdots$$

which defines exact sequences for all cohomology theories after Puppe, Eckmann-Hilton). Hence h' is null-homotopic, $\mathrm{Sq}_{h'}^{q+1} = 0$, and hence $\psi(i^* x) = 0$. □

Proof of Theorem (III.4.8). From (I.2.7) we have an exact sequence (with $\mathbb{Z}_2$ coefficients)

$$\begin{array}{ccccccc} \to K^q(X) & \xrightarrow{i^*} & K^q(Y) & \xrightarrow{\delta} & K^{q+1}(X, Y) \to \\ \downarrow{\scriptstyle \cong} & & \downarrow{\scriptstyle \cong} & & \downarrow{\scriptstyle \cong} \\ \to K_{q+1}(X, Y) & \xrightarrow{\partial} & K_q(Y) & \xrightarrow{i_*} & K_q(X) \longrightarrow \end{array}$$

Also $i^* = \mathrm{Hom}(i_*, \mathbb{Z}_2)$, so $\mathrm{rank}\, i^* K^q(X) = \frac{1}{2} \mathrm{rank}\, K^q(Y)$. Now since $K^q(Y_2) = 0$, it follows that $K^q(Y_1) = K^q(Y)$, and that $\psi_1 = \psi$, where ψ_1 is defined by (f_1, b_1), ψ by (f, b). Then by (III.1.13) and (III.4.13) $c(\psi) = 0$, so $c(\psi_1) = c(f_1, b_1) = 0$. □

Let (f, b), $f: (X, Y) \to (A, B)$ be a normal map of Poincaré pairs, and suppose (X, Y) and (A, B) are sums of Poincaré pairs $X = X_1 \cup X_2$, $A = A_1 \cup A_2$, $X_0 = X_1 \cap X_2$, $A_0 = A_1 \cap A_2$, $Y_i = X_i \cap Y$, $B_i = A_i \cap B$, $f(X_i) \subset A_i$, $(X_i, X_0 \cup Y_i)$, $(A_i, A_0 \cup B_i)$ $i = 1, 2$, are Poincaré pairs oriented compatibly with (X, Y) and (A, B) (see I.3.2). Set

$$f_i = f \mid X_i : (X_i, X_0 \cup Y_i) \to (A_i, A_0 \cup B_i), \qquad i = 1, 2$$

$$f_0 = f \mid X_0 : (X_0, Y_0) \longrightarrow (A_0, B_0),$$

and b_i the appropriate restriction of b.

Now suppose that $(f|Y)^*: H^*(B;\mathbb{Z}_2) \to H^*(Y;\mathbb{Z}_2)$ and

$$f_0^*: H^*(A_0;\mathbb{Z}_2) \to H^*(X_0;\mathbb{Z}_2)$$

are isomorphisms. It follows easily from arguments with the Mayer-Vietoris sequence that $(f_i|X_0 \cup Y_i)^*$, $i=1,2$ are isomorphisms so that $c(f,b)$, $c(f_1,b_1)$ and $c(f_2,b_2)$ are all defined.

III.4.14 Theorem. $c(f,b) = c(f_1,b_1) + c(f_2,b_2)$.

Proof. Let ψ, ψ_1 and ψ_2 be the quadratic forms defined on $K^q(X,Y)$, $K^q(X_1, X_0 \cup Y_1)$ and $K^q(X_2, X_0 \cup Y_2)$ respectively. An argument with the Mayer-Vietoris sequence (which is really the exact sequence of the triple of pairs $(X_0, Y_0) \subset (X, Y) \subset (X, Y \cup X_0)$, where the last pair is replaced by the excisive pair $(X_1, X_0 \cup Y_1) \cup (X_2, X_0 \cup Y_2)$) gives an isomorphism $\varrho_1 + \varrho_2: K^q(X_1, X_0 \cup Y_1) + K^q(X_2, X_0 \cup Y_2) \to K^q(X, Y)$ where ϱ_1 is defined by the diagram

$$\begin{array}{ccc} K^q(X_1, X_0 \cup Y_1) & \xleftarrow{\cong} & K^q(X, X_2 \cup Y) \\ & \searrow^{\varrho_1} & \downarrow \\ & & K^q(X, Y) \end{array}$$

where the isomorphism comes from an excision, and the vertical arrow is induced by inclusion (similar for ϱ_2).

It remains to show:

III.4.15 $\psi(\varrho_i x) = \psi_i(x)$ for $x \in K^q(X_i, X_0 \cup Y_i)$.

Then ψ is isomorphic to the direct sum $\psi_1 + \psi_2$, so that

$$c(\psi) = c(\psi_1) + c(\psi_2)$$

and the theorem will follow.

Consider the diagram:

$$\begin{array}{ccccc} \Sigma^s(A/B) & \xrightarrow{g} & \Sigma^s(X/Y) & & \\ \downarrow \alpha & & \downarrow \beta & & \\ \Sigma^s(A/A_2 \cup B) & \xrightarrow{\bar{g}_1} & \Sigma^s(X/X_2 \cup Y) & & \\ \downarrow \alpha' & & \downarrow \beta' & & \\ \Sigma^s(A_1/A_0 \cup B_1) & \xrightarrow{g_1} & \Sigma^s(X_1/X_0 \cup Y_1) & \xrightarrow{\Sigma^s \varphi} & \Sigma^s K(\mathbb{Z}_2, q) \end{array}$$

where α', β' are homeomorphisms, α, β are the natural collapsing maps $\varrho_1 = (\beta' \beta)^*$, $\varphi^*(\iota) = x \in K^q(X_1, X_0 \cup Y_1)$. The diagram can be shown to be commutative (compare (III.4.11)) and $\alpha' \alpha$ is of degree 1. If $h_1 = (\Sigma^s \varphi) g_1$, then $\psi_1(x) = (\mathrm{Sq}_{h_1}^{q+1}(\Sigma^s(\iota)))(\Sigma^s[A_1])$. Now $\Sigma^s[A_1] = (\alpha'\alpha)_* \Sigma^s[A]$, so that if $h_1' = h_1\alpha'\alpha$, then $\mathrm{Sq}_{h_1}^{q+1}(\Sigma^s(\iota))(\Sigma^s[A_1]) = (\mathrm{Sq}_{h_1'}^{q+1}(\Sigma^s(\iota)))(\Sigma^s[A])$. Now $h_1' = (\Sigma^s\varphi)(\beta'\beta)g$, and since β and β' are s-fold suspensions, it follows that $(\Sigma^s\varphi)(\beta'\beta) = \Sigma^s\varphi'$, $\varphi' : X/Y \to K(\mathbb{Z}_2, q)$, $\varphi'^*(\iota) = \varrho_1(x)$. Hence

$$\psi(\varrho_1 x) = (\mathrm{Sq}_{h_1'}^{q+1}(\Sigma^s(\iota)))\,\Sigma^s[A],$$

and $\psi_1(x) = \psi(\varrho_1 x)$. □

Now suppose (A, B) is a Poincaré complex of dimension m, and ξ is a linear bundle over A, $g : (M, \partial M) \to (A, B)$ is a map of degree 1 and $b : \nu \to \xi$ is a linear bundle map, ν is the normal bundle ν of $(M, \partial M)$ in (D^{m+k}, S^{m+k-1}); i.e. (f, b) is a normal map in the sense of Chapter II. Then by (I.4.19), the enriched Spivak uniqueness theorem, there is a fibre homotopy equivalence (unique up to homotopy) $b' : \xi \to \eta$ such that $T(b')_*(T(b)_*(\alpha)) = \beta$, where

$$\alpha \in \pi_{m+k}(T(\nu), T(\nu \mid \partial M)), \quad \beta \in \pi_{m+k}(T(\eta), T(\eta \mid B))$$

are the natural collapsing maps. Then $(f, b'b)$, $b'b : \nu \to \eta$ is a normal map in the sense of this chapter, and we define

$$\sigma(f, b) = c(f, b'b) \in \mathbb{Z}_2$$

if $m = 4k+2$ and if $(f \mid \partial M)^* : H^*(B; \mathbb{Z}_2) \to H^*(\partial M; \mathbb{Z}_2)$ is an isomorphism.

III.4.16 Proposition. *The value of $\sigma(f, b)$ is independent of the choice of $\beta \in \pi_{m+k}(T(\eta), T(\eta \mid B))$, and thus depends only on the normal map (f, b).*

Proof. Let $\beta_i \in \pi_{m+k}(T(\eta), T(\eta \mid B))$ $i = 1, 2$ be two elements such that $h(\beta_i) \cap U_\eta = [A]$. Then by (I.4.19), there is a fibre homotopy equivalence $e : \eta \to \eta$ such that $T(e)_*(\beta_1) = \beta_2$. If $b_i : \xi \to \eta$ are fibre homotopy equivalences such that

$$T(b_i)_*(\bar{\alpha}) = \beta_i, \quad i = 1, 2 \quad (\bar{\alpha} = T(b)_*(\alpha) \in \pi_{m+k}(T(\xi), T(\xi \mid B))),$$

then b_2 is fibre homotopic to eb_1, by (I.4.19), so $T(b_2) \sim T(e)\,T(b_1)$. It follows that $g_2 \sim g_1 t$, where g_i is S-dual to $T(b_i)$, t is S-dual to $T(e)$, so $t : \Sigma^s A/B \to \Sigma^s A/B$ is a homotopy equivalence. Hence, for the two maps $h_i : \Sigma^s A/B \to \Sigma^s K(\mathbb{Z}_2, q)$, $h_i = (\Sigma^s\varphi)g_i$, $\varphi : M/\partial M \to K(\mathbb{Z}_2, q)$, $h_2 \sim th_1$. Hence

$$\begin{aligned} &\mathrm{Sq}_{h_2}^{q+1}(\Sigma^s(\iota))(\Sigma^s[A]) \\ &\quad = \mathrm{Sq}_{th_1}^{q+1}(\Sigma^s(\iota))(\Sigma^s[A]) = \mathrm{Sq}_{h_1}^{q+1}(\Sigma^s\iota)(t_*\Sigma^s[A]) = \mathrm{Sq}_{h_1}^{q+1}(\Sigma^s\iota)(\Sigma^s[A]), \end{aligned}$$

and the quadratic form ψ is independent of the choice of β. □

§ 5. Product Formulas

In this paragraph we prove the product formula (II.1.7) for σ due to Sullivan. This generalizes the classical formula for the Index.

Let $(f_1, b_1), (f_2, b_2)$ be normal maps, $f_i:(X_i, Y_i)\to(A_i, B_i)$, (X_i, Y_i), (A_i, B_i) Poincaré pairs, $i=1,2$. Suppose $(f_i|Y_i)_*: H_*(Y_i)\to H_*(B_i)$ are isomorphisms. What is the relation between $\sigma(f_1), \sigma(f_2)$ and $\sigma(f_1\times f_2)$? We note that $(f_1\times f_2, b_1\times b_2)$ is a normal map but the boundary of $X_1\times X_2$ is $\partial(X_1\times X_2)=X_1\times Y_2\cup Y_1\times X_2$, so $f_1\times f_2|\partial(X_1\times X_2)$ does not induce homology isomorphism except in special circumstances, (see III.5.6 below).

Note that $(x_1\otimes x_2, y_1\otimes y_2)=(x_1, x_2)(y_1, y_2)$ for

$$x_1, y_1\in H^*(X_1, Y_1), x_2, y_2\in H^*(X_2, Y_2).$$

Hence the bilinear form on $H^*((X_1, Y_1)\times(X_2, Y_2); F)$ is the tensor product of the individual forms.

III.5.1 Lemma. *For the tensor product of bilinear forms on $V_1\otimes V_2$, V_i vector spaces over R,* $\operatorname{sgn}(V_1\otimes V_2)=\operatorname{sgn}(V_1)\operatorname{sgn}(V_2)$.

Proof. We may assume that we have chosen bases $a_1, \ldots, a_k$ for V_1 and $b_1, \ldots, b_l$ for V_2 so that $(a_i, a_j)=(b_i, b_j)=0$ for $i\neq j$, i.e. they are in diagonal form. Then $a_i\otimes b_j$ is a basis for $V_1\otimes V_2$ which puts it in diagonal form. Now if $p_1=$ number of a_i such that $(a_i, a_i)>0$, $n_1=$ number a_i such that $(a_i, a_i)<0$, (similarly p_2, n_2), then the number of $a_i\otimes b_j$ such that $(a_i\otimes b_j, a_i\otimes b_j)>0$ is $p_1p_2+n_1n_2$, since if $(a_i, a_i)(b_j, b_j)>0$, both (a_i, a_i) and (b_j, b_j) are simultaneously $+1$ or -1. Hence while

$$\operatorname{sgn}(V_1)=p_i-n_i, \quad i=1,2,$$

we also have

$$\begin{aligned}\operatorname{sgn}(V_1\otimes V_2)&=p_1p_2+n_1n_2-p_1n_2-n_1p_2=(p_1-n_1)(p_2-n_2)\\&=\operatorname{sgn}(V_1)\operatorname{sgn}(V_2).\quad\square\end{aligned}$$

III.5.2 Lemma. *Let* $\dim X_1=4m$, $\dim X_2=4n$. *The restriction of* $(\,,\,)$ *on* $\Sigma H^i(X_1, Y_1)\otimes H^j(X_2, Y_2)$ *for* $i\neq 2m$, $j\neq 2n$, $i+j=2(m+n)$, *has signature* 0.

Proof. If $a\in H^i(X_1, Y_1)$, $a'\in H^j(X_1, Y_1)$, $i>2m$, $j>2m$, then $aa'\in H^{i+j}(X_1, Y_1)=0$, so $(a, a')=0$ and $(a\otimes b, a'\otimes b')=0$, (similarly for $b, b'\in H^*(X_2, Y_2)$). Now $\Sigma H^i(X_1, Y_1)\otimes H^j(X_2, Y_2)$ $i\neq 2m, i+j=2m+2n$ $=\sum_{i>2m} H^i\otimes H^j+\sum_{j>2n} H^i\otimes H^j$. Hence the first and the second are self-annihilating subspaces. It follows that the signature is zero, (compare with (III.1.2)). $\square$

III.5.3 Proposition. *If* $\dim X_1 = 4m$, $\dim X_2 = 4n$, *then*

$$I(X_1 \times X_2) = I(X_1) I(X_2).$$

Proof.

$$\begin{aligned} H^{2(m+n)}&((X_1, Y_1) \times (X_2, Y_2)) \\ &= \sum_{i \neq 2m} H^i(X_1, Y_1) \otimes H^j(X_2, Y_2) + H^{2m}(X_1, Y_1) \otimes H^{2n}(X_2, Y_2) \end{aligned}$$

as an orthogonal direct sum. By (III.5.2) the signature of the form on the first summand is zero and by (III.5.1) the signature on the second is the product of the signatures. □

III.5.4 Theorem. *Let* $f_i: (X_i, Y_i) \to (A_i, B_i)$, $i = 1, 2$ *be maps of degree* 1 *with* $\dim X_1 = 4m$, $\dim X_2 = 4n$. *Then*

$$I(f_1 \times f_2) = I(f_1) I(A_2) + I(A_1) I(f_2) + I(f_1) I(f_2).$$

Note that this formula together with the relation $\sigma(f) = 8 I(f)$ yields (I.1.7) (i).

Proof. By (III.2.8),

$$I(f_1 \times f_2) = I(X_1 \times X_2) - I(A_1 \times A_2) = I(X_1) I(X_2) - I(A_1) I(A_2),$$

by (III.5.3). Now $I(X_i) = I(A_i) + I(f_i)$, so

$$\begin{aligned} I(f_1 \times f_2) &= (I(A_1) + I(f_1))(I(A_2) + I(f_2)) - I(A_i) I(A_2) \\ &= I(A_1) I(f_2) + I(f_1) I(A_2) + I(f_1) I(f_2). \quad \square \end{aligned}$$

III.5.5 Remark. If $\dim X_1 \times X_2 = 4k$ and $\dim X_i \not\equiv 0(4)$ $i = 1$ or 2, then $I(f_1 \times f_2) = 0$ and (III.5.4) still holds.

Proof. If $\dim X_i$ is odd $i = 1$ or 2, (III.5.2) gives the result. If $\dim X_1 = 2m \equiv 2(4)$ then $(\,,\,)$ on $H^m(X_1, Y_1)$ is skew symmetric so that $(x, x) = 0$ for $x \in H^m(X_1, Y_1; R)$. Hence there is a symplectic basis for $H^m(X_1, Y_1; R)$, $\{a_i, b_i\}$ with $(a_i, a_j) = (b_i, b_j) = 0$, $(a_i, b_j) = \delta_{ij}$. Let A = subspace spanned by the a_i's, B = subspace spanned by the b_i's. Then $A \otimes H^n(X_2, Y_2)$ is a self-annihilating subspace of half the dimension of $H^m(X_1, Y_1) \otimes H^n(X_2, Y_2)$, hence the signature on $H^m(X_1, Y_1) \otimes H^n(X_2, Y_2)$ is zero so the result follows from (III.5.2) as in (III.5.4). □

There remains one case to consider, i.e. what is $\sigma(f_1 \times f_2)$ when the dimension of $X_1 \times X_2$ is $4k+2$ and σ is defined. Namely to define σ in this case it is necessary that $f_1 \times f_2 | \partial(X_1 \times X_2)$ should induce homology isomorphism with $\mathbb{Z}_2$ coefficients.

III.5.6 Lemma. *Suppose* $f_i: (X_i, Y_i) \to (A_i, B_i)$ *are maps of degree* 1, *and suppose* $(f_i | Y_i)_*: H_*(Y_i, G) \to H_*(B_i; G)$ *are isomorphisms* $G = \mathbb{Z}$ *or a field*, $i = 1, 2$. *Then* $f_1 \times f_2 | \partial(X_1 \times X_2)$ *induces an isomorphism*

$$H_*(\partial(X_1 \times X_2); G) \to H_*(\partial(A_1 \times A_2); G)$$

if and only if for each $i=1,2$, *either (i)* $Y_i=\emptyset=B_i$ *or (ii)*

$$f_{i+1}: H_*(X_{i+1}; G) \to H_*(A_{i+1}; G)$$

is an isomorphism, $(i+1=1$ *if* $i=2)$.

Proof. Recall that $\partial(X_1 \times X_2) = X_1 \times Y_2 \cup Y_1 \times X_2$, with

$$Y_1 \times Y_2 = X_1 \times Y_2 \cap Y_1 \times X_2,$$

(similarly for $\partial(A_1 \times A_2)$). Since $(f_i | Y_i)_*: H_*(Y_i; G) \to H_*(B_i; G)$ are isomorphisms $i=1,2$, it follows that

$$(f_1 \times f_2 | Y_1 \times Y_2)_*: H_*(Y_1 \times Y_2; G) \to H_*(B_1 \times B_2; G)$$

is an isomorphism, from the Künneth formula. Consider the map of Mayer-Vietoris sequences induced by $f_1 \times f_2$ on $\partial(X_1 \times X_2)$ into $\partial(A_1 \times A_2)$, and since $f_1 \times f_2$ induces isomorphism on the intersection $Y_1 \times Y_2$ into $B_1 \times B_2$ it follows that

$$\ker(f_1 \times f_2 | \partial(X_1 \times X_2))_* \cong \ker(f_1 \times f_2 | Y_1 \times X_2)_* + \ker(f_1 \times f_2 | X_1 \times Y_2)_* .$$

If $Y_1 \neq \emptyset$ then $1 \otimes \ker f_{2*} \subset \ker(f_1 \times f_2 | Y_1 \times X_2)_*$ and if $Y_2 \neq \emptyset$ then $(\ker f_1)_* \otimes 1 \subset \ker(f_1 \times f_2 | X_1 \times X_2)_*$. Hence if $(f_1 \times f_2 | \partial(X_1 \times X_2))_*$ is an isomorphism, and if $Y_i \neq \emptyset$ then $\ker f_{i+1*} = 0$ and

$$f_{i+1*}: H_*(X_{i+1}; G) \to H_*(A_{i+1}; G)$$

is an isomorphism, since maps of degree 1 are onto in homology.

On the other hand if either $Y_1 = \emptyset$ or f_{2*} is an isomorphism, then either $Y_1 \times X_2$ is empty or $(f_1 \times f_2 | Y_1 \times X_2)_*$ is an isomorphism (similarly for $X_2 \times Y_1$). Hence $(f_1 \times f_2 | \partial(X_1 \times X_2))_*$ is an isomorphism. □

III.5.7 Theorem. *Let* (f_i, b_i) *be normal maps,* $f_i: (X_i, Y_i) \to (A_i, B_i)$ $i=1,2$, *and suppose*

$$(f_1 \times f_2 | \partial(X_1 \times X_2))_*: H_*(\partial(X_1 \times X_2); \mathbb{Z}_2) \to H_*(\partial(A_1 \times A_2); \mathbb{Z}_2)$$

is an isomorphism. Then

$$c(f_1 \times f_2, b_1 \times b_2) = \chi(A_1)\, c(f_2, b_2) + c(f_1, b_1)\, \chi(A_2)$$

where χ *denotes the Euler characteristic.*

This implies (II.1.7) (ii).

Note that if either dimension is odd $c(f_1 \times f_2, b_1 \times b_2)$ is automatically zero. Also (III.5.7) and (III.5.5) together completely determine $\sigma(f_1 \times f_2)$ when it is defined.

The proof proceeds by a sequence of lemmas and takes up the remainder of this section.

Let $g_i : \Sigma^s(A_i/B_i) \to \Sigma^s(X_i/Y_i)$ be the S-duals of $T(b_i): T(\xi_i) \to T(\eta_i)$, $i=1,2$, as at the beginning of § 4, and let

$$g : \Sigma^t(A_1 \times A_2/\partial(A_1 \times A_2)) \to \Sigma^t(X_1 \times X_2/\partial(X_1 \times X_2))$$

be the S-dual of $T(b_1 \times b_2): T(\xi_1 \times \xi_2) \to T(\eta_1 \times \eta_2)$, where ξ_i, η_i are the Spivak normal fibre spaces of X_i, A_i respectively so that $\xi_1 \times \xi_2, \eta_1 \times \eta_2$ are the Spivak normal fibre spaces of $X_1 \times X_2$, $A_1 \times A_2$. Recall that $\partial(A_1 \times A_2) = A_1 \times B_2 \cup B_1 \times A_2$ (similarly for $\partial(X_1 \times X_2)$) so that

$$A_1 \times A_2/\partial(A_1 \times A_2) = (A_1/B_1) \wedge (A_2/B_2)$$

and $X_1 \times X_2/\partial(X_1 \times X_2) = (X_1/Y_1) \wedge (X_2/Y_2)$. If we let s be sufficiently large and $t=2s$ we get:

III.5.8 Lemma. *g is homotopic to $g_1 \wedge g_2$.*

Proof. $T(\xi_1 \times \xi_2) = T(\xi_1) \wedge T(\xi_2)$, $T(\eta_1 \times \eta_2) = T(\eta_1) \wedge T(\eta_2)$ and $T(b_1 \times b_2) = T(b_1) \wedge T(b_2)$. The result then follows from the fact that S-duality preserves $\wedge$ products, (which follows easily from (I.4.14). □

III.5.9 Lemma. *With a field of coefficients,*

$$K^*(f_1 \times f_2) = K^*(f_1) \otimes H^*(X_2, Y_2) + H^*(X_1, Y_1) \otimes K^*(f_2),$$

and

$$K_*(f_1 \times f_2) = K_*(f_1) \otimes H_*(X_2) + H_*(X_1) \otimes K_*(f_2).$$

Proof. This follows from (III.5.8), the fact that K^* is the kernel of $g^*\Sigma^t$, (see III.4.1) and the Künneth formula, using the fact that if $\psi: V \to V'$, $\psi: W \to W'$, V, V', W, W' vector spaces over F, ψ, φ linear maps, then $\ker(\psi \otimes \varphi) = (\ker \psi) \otimes W + V \otimes (\ker \varphi)$.

The proof for K_* is similar and even easier. □

The main point in the proof of (III.5.7) is the following which is a consequence of the Cartan formula.

III.5.10 Proposition. *Let $x \in K^i(f_1; \mathbb{Z}_2)$, $y \in H^j(X_2, Y_2; \mathbb{Z}_2)$, $i+j=n+m$, where* $\dim X_1 = 2n$, $\dim X_2 = 2m$, *$j \leqq m$. Then*

$$\psi(x \otimes y) = \psi(x) \cdot (y, y)$$

so in particular $\psi(x \otimes y) = 0$ if $i > n$, i.e. $j < m$. Similarly $\psi(x \otimes y) = (x, x)\, \psi(y)$ if $x \in H^i(X_1, Y_1; \mathbb{Z}_2)$, $y \in K^j(f_2; \mathbb{Z}_2)$, $i \leqq n$.

Proof. Recall that (see III.4.5),

$$\psi(x \otimes y) = (\mathrm{Sq}_k^{n+m+1}\, (\Sigma^t(\iota_{n+m})))\, \Sigma^t[A_1 \times A_2],$$

where $h = (\Sigma^t \varphi) g$, $\varphi: X_1 \times X_2/\partial(X_1 \times X_2) \to K(\mathbb{Z}_2, n+m)$, $\varphi^*(\iota_{n+m}) = x \otimes y$. Since $X_1 \times X_2/\partial(X_1 \times X_2) = (X_1/Y_1) \wedge (X_2/Y_2)$, φ factors through $\varphi_1 \wedge \varphi_2$, $\varphi_1 : X_1/Y_1 \to K(\mathbb{Z}_2, i)$, $\varphi_2 : X_2/Y_2 \to K(\mathbb{Z}_2, j)$, $\varphi_1^*(\iota_i) = x$, $\varphi_2^*(\iota_j) = y$, $\varphi = \eta(\varphi_1 \wedge \varphi_2)$ where $\eta: K(\mathbb{Z}_2, i) \wedge K(\mathbb{Z}_2, j) \to K(\mathbb{Z}_2, m+n)$, $\eta^*(\iota_{n+m}) = \iota_i \wedge \iota_j$.

By (III.5.8), $g=g_1\wedge g_2$, so that $h=(\Sigma^t\eta)\circ(h_1\wedge h_2)$ where $h_i=(\Sigma^s\varphi_i)g_i$, $t=2s$. We may consider $h_1\wedge h_2=(h_1\wedge 1)\circ(1\wedge h_2)$. Now

$$(h_1\wedge 1)^*(\iota_i\wedge\iota_j)=h_1^*(\iota_i)\wedge\iota_j=0$$

and it follows that $\mathrm{Sq}_{(h_1\wedge 1)}^{n+m+1}(\Sigma^t(\iota_i\wedge\iota_j))$ is defined. From the naturality of functional operations it follows that

III.5.11 $\mathrm{Sq}_h^{n+m+1}(\Sigma^t(\iota_{n+m}))=(1\wedge h_2)^*\,\mathrm{Sq}_{(h_1\wedge 1)}^{n+m+1}(\Sigma^t(\iota_i\wedge\iota_j))$, where the indeterminacy on both sides is zero.

III.5.12 Lemma. *Let $f:S\to T$ be a base point preserving map, $C(f)=T\bigcup_f cS$. Then. $C(f\wedge 1)=(T\wedge Z)\bigcup_{f\wedge 1} c(S\wedge Z)=C(f)\wedge Z$, where $1:Z\to Z$ is the identity.*

Proof. Using the cone with the cone on the base point collapsed to the base point, we have $cS=I\wedge S$, so that $C(f)=T\bigcup_f(I\wedge S)$. Then $C(f)\wedge Z=(T\wedge Z)\cup(I\wedge S\wedge Z)$ where the identifications are by $(1\wedge s\wedge z)\sim(f(s)\wedge z)$, which is exactly by $f\wedge 1$. $\square$

From (III.5.12) it follows that the mapping cone of $h_1\wedge 1$, $C(h_1\wedge 1)=C(h_1)\wedge\Sigma^s K(\mathbb{Z}_2,j)$. If $x'\in H^{s+i-1}(C(h_1);\mathbb{Z}_2)$ such that $\delta x'=\Sigma^s\iota_i\in H^{s+i}(\Sigma^s K(\mathbb{Z}_2,i);\mathbb{Z}_2)$, then

$$\delta(x'\wedge\Sigma^s\iota_j)=\Sigma^t(\iota_i\wedge\iota_j)\in H^{t+i+j}(\Sigma^t K(\mathbb{Z}_2,i)\wedge K(\mathbb{Z}_2,j);\mathbb{Z}_2)$$

so that $\mathrm{Sq}_{h_1\wedge 1}^{k+1}(\Sigma^t(\iota_i\wedge\iota_j))=z$, where

$$(l\wedge 1)^*(z)=\mathrm{Sq}^{k+1}(x'\wedge\Sigma^s\iota_j)\in H^{t+i+j+k}(C(h_1)\wedge\Sigma^s K(\mathbb{Z}_2,j);\mathbb{Z}_2)$$
$$k=m+n=i+j\,,$$

$z\in H^{t+i+j+k}(\Sigma^s(A_1/B_1)\wedge\Sigma^s K(\mathbb{Z}_2,\ j);\ \mathbb{Z}_2)$, $l:C(h_1)\to\Sigma^s A_1/B_1$. By the Cartan formula, $\mathrm{Sq}^{k+1}(x'\wedge\Sigma^s\iota_j)=\sum_{\alpha+\beta=k+1}\mathrm{Sq}^\alpha x'\wedge\mathrm{Sq}^\beta\Sigma^s\iota_j$. If

$$z_\alpha\in H^{s+i-1+\alpha}(\Sigma^s A_1/B_1;\mathbb{Z}_2)$$

is such that $l^*z_\alpha=\mathrm{Sq}^\alpha x'$, then

$$z=\sum_{\alpha+\beta=m+n+1} z_\alpha\wedge\mathrm{Sq}^\beta\iota_j$$

has the right property. Then from (III.5.11) we get that

III.5.13
$$\begin{aligned}\psi(x\otimes y)&=\mathrm{Sq}_h^{n+m+1}(\Sigma^t(\iota_{n+m}))\Sigma^t[A_1\times A_2]\\&=(1\wedge h_2)^*(\Sigma z_\alpha\wedge\Sigma^s\mathrm{Sq}^\beta\iota_j)\Sigma^t[A_1\times A_2]\\&=(\Sigma z_\alpha\wedge h_2^*\Sigma^s\mathrm{Sq}^\beta\iota_j)(\Sigma^s[A_1]\wedge\Sigma^s[A_2])\\&=\Sigma z_\alpha(\Sigma^s[A_1])\cdot(h_2^*\Sigma^s\mathrm{Sq}^\beta\iota_j)\Sigma^s[A_2]\,.\end{aligned}$$

Now if $\dim h_2^* \Sigma^s \mathrm{Sq}^\beta \iota_j < 2m+s = \dim \Sigma^s[A_2]$, or $\beta < 2m-j$, then $(h_2^* \Sigma^s \mathrm{Sq}^\beta \iota_j)(\Sigma^s[A_2]) = 0$, and on the other hand

$$\mathrm{Sq}^\beta \iota_j = \begin{cases} 0 & \text{if } \beta > j \\ \iota_j^2 & \text{if } \beta = j. \end{cases}$$

Since $j \leqq m$, it follows that for $\beta = 2m-j$, $\beta \geqq m \geqq j \geqq \beta$ and so the only non-zero term in (III.5.13) occurs when $\beta = m = j$, so that $\psi(x \otimes y) = 0$ if $j < m$. If $j = m = \beta$, then $\alpha = n+1$ so that $z_\alpha = \mathrm{Sq}_{h_1}^{n+1}(\Sigma^s \iota_n)$ and from (III.5.13),

III.5.14
$$\begin{aligned}\psi(x \otimes y) &= (\mathrm{Sq}_{h_1}^{n+1}(\Sigma^s \iota_n)\Sigma^s[A_1]) \cdot (h_2^* \Sigma^s(\iota_m)^2)(\Sigma^s[A_2]) \\ &= \psi(x) \cdot ((h_2^* \Sigma^s(\iota_m)^2)(\Sigma^s[A_2])).\end{aligned}$$

Since $h_2 = (\Sigma^s \varphi_2) g_2$ so

$$h_2^* \Sigma^s(\iota_m^2) = g_2^* \Sigma^s \varphi_2^* \Sigma^s(\iota_m^2) = g_2^* \Sigma^s \varphi_2^*(\iota_m^2) = g_2^* \Sigma^s y^2 .$$

Then (III.5.14) becomes

$$\begin{aligned}\psi(x \otimes y) &= \psi(x)\,(g_2^* \Sigma^s(y^2))\,(\Sigma^s[A_2]) \\ &= \psi(x)\,(\Sigma^s(y^2) g_{2*} \Sigma^s[A_2]) \\ &\doteq \psi(x)\,((\Sigma^s(y^2))\,(\Sigma^s \alpha_{2*}[A_2])) \\ &= \psi(x)\,(y^2[X_2]) = \psi(x) \cdot (y, y)\end{aligned}$$

using (III.4.1), and the fact $\alpha_{2*}[A_2] = [X_2]$. This completes the proof of (III.5.10). □

By (I.2.5) and (I.2.9), we have an orthogonal splitting

$$H^*(X_i, Y_i) = K^*(f_i) + f_i^* H^*(A_i, B_i),$$

so that

III.5.15
$$\begin{aligned}K^*(f_1 \times f_2) = K^*(f_1) \otimes K^*(f_2) &+ K^*(f_1) \otimes f_2^* H^*(X_2, Y_2) \\ &+ f_1^* H^*(X_1, Y_1) \otimes K^*(f_2).\end{aligned}$$

and this is an orthogonal splitting.

Now recall that if $x \in K^n(f_1)$, $(x, x) = 0$ (since $(x, x) = 2\psi(x) = 0$ by (III.4.6), or $(x, x) = (x, v_n(X_1)) = (x, f^* v_n(A_1)) = 0$ by (III.3.8)). Hence if $x \otimes y \in K^*(f_1) \otimes K^*(f_2)$, then $\psi(x \otimes y) = 0$, by (III.5.10). Such $x \otimes y$ form a basis for $K^*(f_1) \otimes K^*(f_2)$ so that the Arf invariant of $\psi | K^*(f_1) \otimes K^*(f_2)$ is zero and we have

III.5.16 $c(\psi) = c(\psi | K^*(f_1) \otimes \mathrm{im} f_2^*) + c(\psi | \mathrm{im} f_1^* \otimes K^*(f_2))$.

Note that we have an orthogonal direct sum:

$$\begin{aligned}(K^*(f_1) \otimes \mathrm{im} f_2^*)^{n+m} &= K^n(f_1) \otimes f_2^* H^m(A_2, B_2) \\ &+ \sum_{\substack{i \neq n \\ i+j=n+m}} K^i(f_1) \otimes f_2^* H^j(A_2, B_2)\end{aligned}$$

Now $\psi | \sum_{\substack{i>n \\ i+j=n+m}} K^i(f_1) \otimes H^j(A_2, B_2) \equiv 0$ by (III.5.10) and by Poincaré duality this is a subspace of half the rank of $\sum_{i \neq n} K^i(f_1) \otimes H^j(A_2, B_2)$, so that the Arf invariant of ψ on this space is zero. Similar reasoning applies to $\operatorname{im} f_1^* \otimes K^*(f_2)$ and we get:

III.5.17

$$c(\psi) = c(\psi | K^n(f_1) \otimes f_2^* H^m(A_2, B_2)) + c(\psi | f_1^* H^n(A_1, B_1) \otimes K^m(f_2)).$$

Let V_1 be a one dimensional vector space over $\mathbb{Z}_2$ with bilinear form $(c, c) = 1$ where c is the basis element of V_1. Let V_0 be a two dimensional vector space over $\mathbb{Z}_2$ with basis $\{a, b\}$, and bilinear form $(a, a) = (b, b) = 0$, $(a, b) = (b, a) = 1$.

III.5.18 Lemma. *Let $(,): V \otimes V \to \mathbb{Z}_2$ be a non-singular symmetric bilinear form over $\mathbb{Z}_2$. Then $V \cong \varepsilon V_1 + k V_0$, where $\varepsilon = 0$, 1 or 2, $2k + \varepsilon = \dim V$.*

Proof. Consider the map $\varphi: V \to \mathbb{Z}_2$, $\varphi(x) = (x, x)$. This is linear since

$$\begin{aligned} \varphi(x+y) &= (x+y, x+y) = (x, x) + (y, x) + (x, y) + (y, y) \\ &= (x, x) + (y, y) + 2(x, y) = \varphi(x) + \varphi(y). \end{aligned}$$

If $\varphi \equiv 0$, the $(,)$ is sympletic and non-singular, so that there is a symplectic basis which gives an isomorphism $V \cong k V_0$.

Suppose $\varphi \neq 0$. Since $(,)$ is non-singular there is a unique element $v \in V$ such that $\varphi(x) = (x, v)$ for all $x \in V$. Let $W = \ker \varphi \subset V$, $T =$ subspace generated by $v \subset V$. We consider two cases:

Case 1. $\varphi(v) = (v, v) = 1$, i.e. $v \notin W$. Then $V = W + T$, W and T are orthogonal, $(,)$ is non-singular and symplectic on W so $W \cong k V_0$, and $T \cong V_1$. Hence $V \cong V_1 + k V_0$.

Case 2. $\varphi(v) = (v, v) = 0$, i.e. $v \in W$. Let $u \in V$ such that $\varphi(u) = (u, v) = 1$, so that if S is generated by u, $V = W + S$. Let

$$R = (\text{annihilator of } u) \cap W = \{y \in V \text{ such that } (y, u) = (y, v) = 0\}.$$

Then $W = R + T$ so that $V = R + T + S$ and R is orthogonal to $T + S$. Now $(,)$ is symplectic and non-singular on R, so $R \cong k V_0$. In $T + S$, we have $(u, u) = \varphi(u) = (u, v) = 1$, $(v, v) = \varphi(v) = 0$. Then

$$(u+v, u+v) = (u, u) + (v, v) = (u, u) = 1, (u, u+v) = (u, u) + (u, v) = 0$$

so that with the basis $\{u, u+v\}$, $T + S \cong 2V_1$. □

III.5.19 Remark. Note that $\varepsilon_1 V_1 + k_1 V_0 \cong \varepsilon_2 V_1 + k_2 V_0$, $0 \leqq \varepsilon_i \leqq 2$, if and only if $\varepsilon_1 = \varepsilon_2$ and $k_1 = k_2$, so that the decomposition of (III.5.18) is uniquely determined.

Let U, V be $\mathbb{Z}_2$ vector spaces, $q: U \to \mathbb{Z}_2$ a quadratic form with $\langle\, , \rangle: U \otimes U \to \mathbb{Z}_2$ as associated bilinear form, $(\, ,): V \otimes V \to \mathbb{Z}_2$ a bilinear form. On $U \otimes V$ we may define a quadratic form $\psi: U \otimes V \to \mathbb{Z}_2$ by defining it on basis elements by $\psi(x \otimes y) = q(x) \cdot (y, y)$, (compare III.5.10)), so that ψ has $\langle\, , \rangle \cdot (\, ,)$ as associated bilinear form.

Let U_0, U_1 be two dimensional vector spaces over $\mathbb{Z}_2$ with quadratic forms q_0, q_1 respectively, with $c(q_0) = 0, c(q_1) = 1$, (see III.1.5), and let V_0, V_1 be as above.

III.5.20 Lemma. *$U_i \otimes V_1 \cong U_i$, $U_i \otimes V_0 \cong 2U_0$, $i = 0, 1$, as spaces with quadratic forms.*

Proof. Calculate on bases of the various spaces, and use (III.1.6). □

Proof of (III.5.7). By definition

$$
\begin{aligned}
c(f_1 \times f_2, b_1 \times b_2) &= c(\psi \text{ on } K^*(f_1 \times f_2)) \\
&= c(\psi \,|\, K^n(f_1) \otimes f_2^* H^m(A_2, B_2)) + c(\psi \,|\, f_1^* H^n(A_2, B_1) \otimes K^m(f_2))
\end{aligned}
$$

by (III.5.17). By (III.5.10), ψ on these two spaces is the tensor product of the quadratic form on the K^* factor and the bilinear form on the im f_i^* factor. Let $c_i = c(f_i, b_i)$ so that

$$K^n(f_1) \cong c_1 U_1 + t_1 U_0, \quad K^m(f_2) \cong c_2 U_1 + t_2 U_0 .$$

Since $c(f_1 \times f_2, b_1 \times b_2)$ is defined by hypothesis, so that

$$(f_1 \times f_2 \,|\, \partial(X_1 \times X_2))_*$$

is an isomorphism with $\mathbb{Z}_2$ coefficients, it follows from (III.5.6) that for each i, either a) $Y_i = B_i = \emptyset$ or b) $K^*(f_{i+1}) = 0$. In case b), the appropriate term, (say if $i = 1$, im $f_1^* \otimes K^*(f_2)$) is zero, so we may assume that for each non-zero term a) holds, i.e. $B_i = \emptyset$ and hence the bilinear forms on $H^n(A_1, B_1) = H^n(A_1)$, $H^m(A_2, B_2) = H^m(A_2)$ are non-singular.

Let $H^n(A_1) \cong \varepsilon_1 V_1 + k_2 V_0$. Then by (III.5.20), we have that

$$
\begin{aligned}
K^n(f_1) \otimes f_2^* H^m(A_2) &\cong (c_1 U_1 + t_1 U_0) \otimes (\varepsilon_2 V_1 + k_2 V_0) \\
&\cong c_1 \varepsilon_2 U_1 + (t_1 \varepsilon_2 + 2c_1 k_2 + 2t_1 k_2) U_0 .
\end{aligned}
$$

A similar argument shows that

$$f_1^* H^n(A_1) \otimes K^m(f_2) \cong c_2 \varepsilon_1 U_1 + (t_2 \varepsilon_1 + 2c_2 k_1 + 2t_2 k_1) U_0$$

so we get

$$c(\psi \mid K^n(f_1) \otimes f_2^* H^m(A_2)) = c_1 \varepsilon_2$$

$$c(\psi \mid f_1^* H^n(A_1) \otimes K^m(f_2)) = c_2 \varepsilon_1, \quad \text{and} \quad c(\psi) = c_1 \varepsilon_2 + c_2 \varepsilon_1 .$$

Now by Poincaré duality rank $H^{n-j}(A_1) = \text{rank}\, H^{n+j}(A_1)$ so that mod 2 $\chi(A_1) = \text{rank}\, H^n(A_1) = \varepsilon_1$ and similarly $\chi(A_2) = \varepsilon_2 \bmod 2$ which completes the proof of (III.5.7). □

Since rank $K_n(f_1) = 0 \bmod 2$ and $H_*(X_1) = K_*(f_1) + \alpha H_*(A_1)$ it follows that $\chi(X_1) = \chi(A_1) = \varepsilon_1 \bmod 2$ and similarly $\chi(X_2) = \chi(A_2) = \varepsilon_2$ $\cdot \bmod 2$.

IV. Surgery and the Fundamental Theorem

In this chapter we develop the techniques of surgery for constructing normal cobordisms and use them to prove the Fundamental Theorem.

The ideas of surgery have their origins in the theory of 2-manifolds, in the process of "cutting off handles", and in general, in the theory of Marston Morse of non-degenerate critical points of differentiable functions. A good modern exposition of the Morse Theory and the applications due to Smale of it to study of differentiable manifolds has been given in the two books of Milnor, [41] and [42].

§ 1. Elementary Surgery and the Group $SO(n)$

We now describe the surgery process on a given smooth manifold M^m.

Suppose $\varphi: S^p \times D^{q+1} \to M^m$, $p+q+1=m$, is a differentiable embedding, into the interior of M if $\partial M \neq \emptyset$. Let $M_0 = M -$ interior $\varphi(S^p \times D^{q+1})$. Then $\partial M_0 = \partial M \cup \varphi(S^p \times S^q)$. We define $M' = M_0 \cup D^{p+1} \times S^q$, with $\varphi(x, y)$ identified to $(x, y) \in S^p \times S^q \subset D^{p+1} \times S^q$. Then M' is a manifold, $\partial M' = \partial M$, and we refer to M' as being the result of doing a surgery using φ, on M. Further, we may define a cobordism W_φ^{m+1} between M and M' as follows: $W_\varphi = M \times [0, 1] \cup (D^{p+1} \times D^{q+1})$ with the identification $(x, y) \in S^p \times D^{q+1} \subset \partial(D^{p+1} \times D^{q+1})$ is identified with $(\varphi(x, y), 1) \subset M \times 1$. Clearly $\partial W_\varphi = M \cup (\partial M \times I) \cup M'$ and we call it the *trace* of the surgery. Unfortunately W_φ is not a smooth manifold with boundary as it stands, but has "corners," i.e. points such as in $\varphi(S^p \times S^q)$, $\partial M \times 0$ and $\partial M \times 1$, where the coordinate neighborhoods naturally look like one quadrant of the plane times R^{m-1}, instead of a Euclidean half space. However, there is a canonical way to make it a smooth manifold with boundary, a process called "straightening the angles" which is described for example in [18, Chapter I].

If W^{m+1} is a manifold with $\partial W = M \cup (\partial M \times I) \cup M'$ and W' has $\partial W' = M' \cup (\partial M' \times I) \cup M''$, then we may define the sum of the two cobordisms by taking $\overline{W} = W \cup W'$ and identifying $M' \subset \partial W$ with $M' \subset \partial W''$. Then it is clear that $\partial \overline{W} = M \cup (\partial M \times I) \cup M''$.

IV.1.1 Theorem. *Let W be a cobordism with $\partial W = M \cup (\partial M \times I) \cup M'$. Then there is a sequence of surgeries based on embeddings φ_i, $i = 1, \ldots, k$ each surgery being on the manifold which results from the previous surgery, and such that W is the sum of $W_{\varphi_1}, \ldots, W_{\varphi_k}$.*

The proof is an immediate consequence of the Morse Lemma, and we refer to [42] for a proof.

IV.1.2 Proposition. *If M' is the result of a surgery on M based on an embedding $\varphi: S^p \times D^{q+1} \to M$, then M is the result of a surgery on M' based on an embedding $\psi: S^q \times D^{p+1} \to M'$ and the traces of the two surgeries are the same.*

Proof. Let W_φ be the trace of φ so that $W_\varphi = M \times I \bigcup_\varphi D^{p+1} \times D^{q+1}$. If we set $M_0 = M -$ interior $\varphi(S^p \times D^{q+1})$, then we may equally well view W_φ as $M_0 \times I \cup (S^p \times D^{q+1} \times I) \cup D^{p+1} \times D^{q+1}$, and $S^p \times D^{q+1} \times I \cup D^{p+1} \times D^{q+1}$ may be reparametrized to be $D^{p+1} \times D^{q+1}$. Thus we may view W_φ as $M_0 \times I \cup D^{p+1} \times D^{q+1}$, united along $S^p \times S^q \times I$. Now from the obviously symmetrical nature of this description, the proposition follows. □

IV.1.3 Proposition. *Let $\varphi: S^p \times D^{q+1} \to M^m$ be a smooth embedding in the interior of M, $p+q+1=m$, and let W_φ be the trace of the surgery based on φ. Then W_φ has $M \bigcup_{\bar{\varphi}} D^{p+1}$ as a deformation retract, where $\bar{\varphi} = \varphi \mid S^p \times 0$.*

Proof. $W_\varphi = (M \times I) \bigcup_\varphi (D^{p+1} \times D^{q+1})$, image $\varphi \subset M \times 1$, so we may deform $M \times I$ to $M \times 1$ leaving $M \times 1 \bigcup_\varphi (D^{p+1} \times D^{q+1})$ fixed. Then $D^{p+1} \times D^{q+1}$ may be deformed onto $(D^{p+1} \times 0) \cup (S^p \times D^{q+1})$, leaving the subspace fixed. This then yields the deformation retraction of W_φ to $M \bigcup_{\bar{\varphi}} D^{p+1}$. □

IV.1.4 Proposition. (a) *Let $f: (M, \partial M) \to (A, B)$ be a map, M an oriented smooth m-manifold, (A, B) a pair of spaces, and let $\varphi: S^p \times D^{q+1} \to \text{interior}\, M$ be a smooth embedding, $p+q+1=m$. Then f extends to $F: (W_\varphi, \partial M \times I) \to (A \times I, B \times I)$ to get a cobordism of f if and only if $f \circ \bar{\varphi}$ is homotopic to the constant map $S^p \to A$.*

(b) *Suppose in addition that η^k is a linear k-plane bundle over A, $b: \nu^k \to \eta^k$ is a linear bundle map covering f, $\nu =$ normal bundle of $(M, \partial M) \subset (D^{m+k}, S^{m+k-1})$, $k >> m$. Then b extends to $\bar{b}: \omega \to \eta$ covering F, where $\omega =$ normal bundle of $W_\varphi \subset D^{n+k} \times I$ if and only if $b \mid (\nu \mid \varphi(S^p))$ extends to $\omega \mid (D^{p+1} \times 0)$, covering $F \mid D^{p+1} \times 0$.*

Proof. Since $M \bigcup_{\bar{\varphi}} D^{p+1}$ is a deformation retract of W_φ, it follows that f extends to W_φ if and only if f extends to $M \bigcup_{\bar{\varphi}} D^{p+1}$. But the latter is true if and only if $f \circ \bar{\varphi}$ is null-homotopic, which proves (a).

For (b), it follows from the bundle covering homotopy property that since $M \bigcup_{\bar{\varphi}} D^{p+1}$ is a deformation retract of W_φ, b extends to ω if and only if b extends to $\omega|(D^{p+1} \times 0)$. □

If (f, b) is a normal map (see Chapter II), $\varphi: S^p \times D^{q+1} \to \operatorname{Interior} M^m$, $m = p+q+1$, $f: (M, \partial M) \to (A, B)$, and if the trace of φ can be made a normal cobordism by extending f and b over W_φ, we will say that the surgery based on φ is a *normal* surgery on (f, b).

From (IV.1.1) we may deduce easily that any normal cobordism rel B is the composite of normal surgeries.

We are here principally interested in normal surgery as a method of constructing normal cobordisms, rather than vice versa.

Let $\varphi: S^p \times D^{q+1} \to \operatorname{Interior} M^m$ be an embedding $m = p+q+1$, W_φ the trace, and M' the result of the corresponding surgery. Now we will discuss the effect of surgery on the homotopy of M, namely the relation between the homotopy groups of M and M', below the "middle dimension."

IV.1.5 Theorem. *If* $p < \dfrac{m-1}{2}$ *then* $\pi_i(M') \cong \pi_i(M)$ *for* $i < p$, *and* $\pi_p(M') \cong \pi_p(M)/\{\bar{\varphi}_* \pi_p(S^p)\}$, *where* $\{X\}$ *denotes the* $\mathbb{Z}[\pi_1(M)]$ *submodule of* $\pi_p(M)$ *generated by* X.

Proof. By (IV.1.3), W_φ is of the same homotopy type as $M \bigcup_{\bar{\varphi}} D^{p+1}$. Hence $\pi_i(W_\varphi) = \pi_i(M)$ for $i < p$, and $\pi_p(W_\varphi) = \pi_p(M)/\{\bar{\varphi}_* \pi_p(S^p)\}$. By (IV.1.2) and (IV.1.3), we have also that $W_\varphi = W_\psi = M' \bigcup_{\bar{\psi}} D^{q+1}$, where $\psi: S^q \times D^{p+1} \to M'$ gives the surgery which makes M' back into M. Hence $\pi_i(W_\varphi) = \pi_i(M')$ for $i < q$, and $\pi_q(W_\varphi) = \pi_q(M')/\{\bar{\psi}_* \pi_q(S^q)\}$. Since $p < \dfrac{m-1}{2}$, then $q > p$, so $\pi_i(M') = \pi_i(W_\varphi)$ for $i \leqq p$ and the result follows. □

The analysis for p near $\dfrac{m}{2}$ is much harder, and will be dealt with in later sections in the 1-connected case.

Let (f, b) be such that $f: (M, \partial M) \to (A, B)$, $b: \nu^k \to \eta^k$, $k >> m$, η a linear bundle over A, $\nu =$ normal bundle of $(M, \partial M) \subset (D^{m+k}, S^{m+k-1})$, and let $\bar{\varphi}: S^p \to \operatorname{Interior} M$, be a smooth embedding. Suppose that f extends to $\bar{F}: \bar{M} \to A$ where $\bar{M} = M \bigcup_{\bar{\varphi}} D^{p+1}$. We consider the problem of "thickening $\bar{M}$ to a normal cobordism" i.e. of extending $\bar{\varphi}$ to a smooth embedding $\varphi: S^p \times D^{q+1} \to \text{Interior}\, M^n$, $m=p+q+1$ such that $\bar{\varphi}=\varphi | S^p \times 0$, and so that $F:(W_\varphi, \partial M \times I) \to (A \times I, B \times I)$ can be covered by a bundle map $\bar{b}: \omega \to \eta$ extending b, where ω is the normal bundle of W_φ in $D^{m+k} \times I$, F is the extension of $\bar{F}$, unique up to homotopy.

IV.1.6 Theorem. *There is an obstruction $\mathcal{O} \in \pi_p(V_{k,q+1})$ such that $\mathcal{O}=0$ if and only if $\bar{\varphi}$ extends to φ such that $F: W_\varphi \to A$ can be covered by $\bar{b}: \omega \to \eta$ extending b as above.*

Here $V_{k,q+1}$ is the space of orthonormal k-frames in R^{k+q+1}.

Proof. If we consider $M \subset D^{m+k}$, since k is very large, we may extend the embedding to $M \bigcup_{\bar{\varphi}} D^{p+1} \subset D^{m+k} \times I$, with D^{p+1} coming in orthogonally to $D^{m+k} \times 0$, and D^{p+1} smoothly embedded. The normal bundle γ of D^{p+1} in $D^{m+k} \times I$ is trivial, i.e. $D^{p+1} \times R^{q+k+1} =$ total space of γ.

Now $\bar{F}$ defines a homotopy of $f\bar{\varphi}$ to a point, which is covered by a bundle homotopy b on $v | \bar{\varphi}(S^p)$, ending with a map of $v | \bar{\varphi}(S^p)$ into a single fibre of η, i.e. a trivialization of $v | \bar{\varphi}(S^p)$, which is well defined up to homotopy. This trivialization of $v | \bar{\varphi}(S^p)$, which is a subbundle of $\gamma | \bar{\varphi}(S^p)$ which is also trivial, therefore defines a map α of S^p into the k-frames in R^{q+k+1}, $\alpha: S^p \to V_{k,q+1}$ and thus defines an element $\mathcal{O} \in \pi_p(V_{k,q+1})$. Now if $\bar{\varphi}$ extends to φ and b extends to $\bar{b}$ as above, then the normal bundle ω of W_φ restricted to D^{p+1}, $\omega | D^{p+1}$ is a subbundle of γ extending $v | \bar{\varphi}(S^p)$, and $\bar{b}$ defines an extension of α to $\alpha': D^{p+1} \to V_{k,q+1}$. Hence $\mathcal{O}=0$ in $\pi_p(V_{k,q+1})$.

Conversely if $\mathcal{O}=0$, then α extends to $\alpha': D^{p+1} \to V_{k,q+1}$, and α' defines a trivial subbundle ω' of dimension k in γ, extending $v | \bar{\varphi}(S^p)$. The subbundle ω'' orthogonal to ω' in γ is trivial (being a bundle over D^{p+1}) and the total space of ω'' is $D^{p+1} \times R^{q+1} \subset D^{p+1} \times R^{q+k+1} =$ total space of γ. Since $\omega'' | \bar{\varphi}(S^p) =$ the normal bundle of $\bar{\varphi}(S^p)$ in M, this embedding defines $\varphi: S^p \times R^{q+1} \subset M$, and α' defines the extension of b to $\bar{b}: \omega \to \eta$, where by construction $\omega | D^{p+1} = \omega'$. □

Now we shall study $V_{k,q+1}$, in order to analyze the obstruction $\mathcal{O}$, (see [60]).

Recall that the group $SO(k+q+1)$ acts transitively on the set of orthonormal k-frames in R^{k+q+1} and $SO(q+1)$ is the subgroup leaving a given frame fixed. Hence $V_{k,q+1} = SO(k+q+1)/SO(q+1)$ and $V_{k,q+1}$ is topologized to make this a homeomorphism, (see [60] or [32]). Further, we recall (see [60] or [32]) that $SO(n) \xrightarrow{i} SO(n+1) \xrightarrow{p} S^n$ is a fibre bundle map, where p is the map which evaluates an orthogonal transformation at the first unit vector, i.e. $p(T) = T(1, 0, \ldots, 0)$, $T \in SO(n+1)$, $(1, 0, \ldots, 0) \in S^n \subset R^{n+1}$.

IV.1.7 Lemma. *$i_*: \pi_i(SO(n)) \to \pi_i(SO(n+1))$ is an isomorphism for $i < n-1$, onto for $i \leqq n-1$.*

Proof. $\pi_i(S^n)=0$ for $i<n$, so the result follows from the exact homotopy sequence:

$$\cdots \longrightarrow \pi_{i+1}(S^n) \xrightarrow{\partial} \pi_i(SO(n)) \xrightarrow{i_*} \pi_i(SO(n+1)) \xrightarrow{p_*} \pi_i(S^n) \longrightarrow \cdots .$$

IV.1.8 Lemma. *The map $p: SO(n+1) \to S^n$ is the projection of the principal $SO(n)$ bundle associated with the oriented tangent bundle of S^n.*

Proof. Let $f=(f_1,\ldots,f_n)$ be a tangent frame to S^n at $v_0=(1,0,\ldots,0)\in S^n$. Define a map $e: SO(n+1) \to F =$ bundle of frames of S^n, $e(T)=$ frame $T(f_1),\ldots,T(f_n)$ at $T(v_0)\in S^n$. Then e is onto, and it is obviously $1-1$. Hence e is a homeomorphism and the lemma follows. □

IV.1.9 Lemma. *The composite $\pi_n(S^n) \xrightarrow{\partial} \pi_{n-1}(SO(n)) \xrightarrow{p_*} \pi_{n-1}(S^{n-1})$ is the boundary in the exact sequence of the tangent S^{n-1} bundle to S^n and is $=0$ if n is odd, multiplication by 2 if n is even.*

Proof. The tangent S^{n-1} bundle is obtained from the bundle of frames by taking the quotient by $SO(n-1)\subset SO(n)=$ the group of the bundle. Hence, we have the commutative diagram

$$\begin{array}{ccc} SO(n) & \xrightarrow{p} & SO(n)/SO(n-1)=S^{n-1} \\ \downarrow{\scriptstyle i} & & \downarrow \\ SO(n+1) & \longrightarrow & SO(n+1)/SO(n-1) \\ \downarrow & & \downarrow{\scriptstyle \bar{p}} \\ S^n & \xrightarrow{\text{Identity}} & S^n . \end{array}$$

It follows that in the exact sequence for the right hand bundle

$$\bar{\partial} = p_* \partial : \pi_i(S^n) \longrightarrow \pi_{i-1}(S^{n-1}) .$$

Now by the Euler-Poincaré Theorem the tangent sphere bundle has a cross-section (there is a non-singular tangent vector field) if and only if the Euler characteristic $\chi(M)=0$. More precisely, the only obstruction to a cross-section to the tangent sphere bundle of a manifold M^m is $\chi(M)g$, where $g\in H^m(M;\mathbb{Z})$ is the class dual to the orientation class of M, (see [32]). Now in case $M=S^n$, the obstruction to a cross-section can also be identified with the "characteristic map" (see [60, (23.4)]) $\bar{\partial}: \pi_n(S^n) \to \pi_{n-1}(S^{n-1})$. Hence $\bar{\partial}=0$ if n is odd, multiplication by 2 if n is even. □

IV.1.10 Theorem. $p_*: \pi_n(SO(n+1)) \to \pi_n(S^n)$ *is onto if and only if* $n=1, 3$ *or* 7.

Proof. If p_* is onto, then there is a map $\alpha: S^n \to SO(n+1)$ such that $p\alpha \sim 1$, and hence the principal bundle of τ_{S^n} has a section and is therefore trivial, i.e. S^n is parallelizable. But it is known (see [36], [5]) that S^n is parallelizable if and only if $n=1$, 3 or 7. □

IV.1.11 Corollary. *kernel* $i_*:\pi_{n-1}(SO(n))\to\pi_{n-1}(SO(n+1))$, *is* $\mathbb{Z}$ *if n is even,* $\mathbb{Z}_2$ *if n is odd and* $n\neq 1,3,7$, *and* 0 *if* $n=1,3$ *or* 7.

Proof. kernel $i_*=\partial\pi_n(S^n)\cong\pi_n(S^n)/p_*\pi_n(SO(n+1))$. If n is odd, by (IV.1.9), $p_*\pi_n(SO(n+1))\supset 2\pi_n(S^n)$, and by (IV.1.10) is not the whole group, if $n\neq 1$, 3 or 7, hence $\pi_n(S^n)/p_*\pi_n(SO(n+1))=\mathbb{Z}_2$ if n is odd, $n\neq 1$, 3 or 7. If $n=1$, 3, or 7, p_* is onto, so kernel $i_*=0$.

If n is even, by (IV.1.9) $p_*\partial$ is a monomorphism, so

$$\partial:\pi_n(S^n)\to\pi_{n-1}(SO(n))$$

is a monomorphism, so kernel $i_*\cong\mathbb{Z}$. □

IV.1.12 Theorem. $\pi_i(V_{k,m})=0$ *for* $i<m$, $\pi_m(V_{k,m})=\mathbb{Z}_2$ *if m is odd,* $\mathbb{Z}$ *if m is even,* $k\geqq 2$. *Further* $j_*:\pi_i(V_{k,m})\to\pi_i(V_{k+1,m})$ *is an isomorphism for* $i\leqq m$, $k\geqq 2$, *and* $j_*:\pi_m(V_{1,m})=\pi_m(S^m)\to\pi_m(V_{k,m})$ *is onto, and an isomorphism if m is even, where j is inclusion.*

Proof. Take $k=2$ so that $V_{2,m}=SO(m+2)/SO(m)$ and we have a natural fibration over $S^{m+1}=SO(m+2)/SO(m+1)$ with fibre $S^m=SO(m+1)/SO(m)$. Also we have a commutative diagram of fibre bundles:

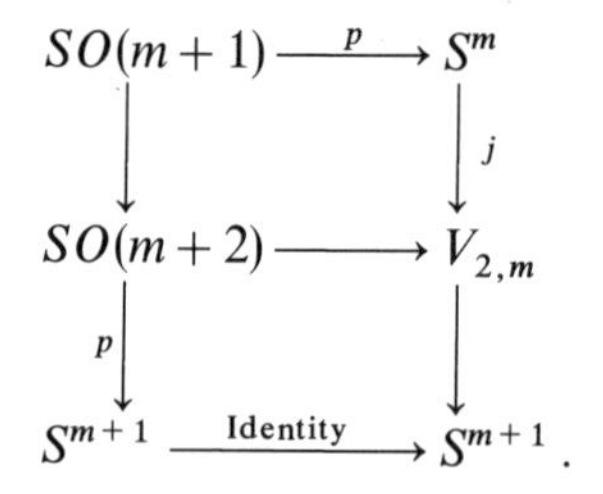

It follows that we have a commutative diagram

$$\begin{array}{ccc}\pi_{m+1}(S^{m+1}) & \xrightarrow{1} & \pi_{m+1}(S^{m+1})\\ \downarrow{\scriptstyle\partial} & & \downarrow{\scriptstyle\partial'}\\ \pi_m(SO(m+1)) & \xrightarrow{p_*} & \pi_m(S^m).\end{array}$$

By (IV.1.9) $p_*\partial=0$ if m is even, $p_*\partial=$ multiplication by 2 if m is odd. Hence $\partial'=p_*\partial$, and from the exact homotopy sequence of the fibre bundle

$$\pi_{i+1}(S^{m+1})\xrightarrow{\partial'}\pi_i(S^m)\xrightarrow{j_*}\pi_i(V_{2,m})\longrightarrow\pi_i(S^{m+1})=0$$

for $i\leqq m$, we obtain j_* is onto for $i\leqq m$, and $\pi_i(V_{2,m})=0$ for $i<m$, $\pi_m(V_{2,m})=\mathbb{Z}$ if m is even, and $\pi_m(V_{2,m})=\mathbb{Z}_2$ if m is odd.

Consider the natural inclusion $V_{k,m}\to V_{k+1,m}$ given by including $SO(m+k)\to SO(m+k+1)$, so that the $SO(m)$ subgroup is preserved.

Then we get the commutative diagram:

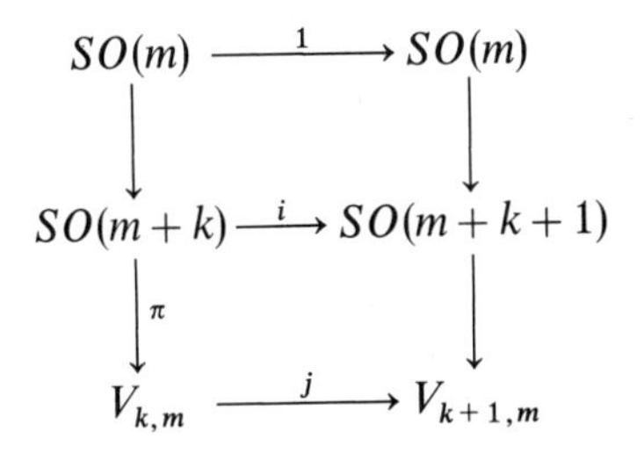

and a corresponding map of exact sequences

$$\begin{array}{ccccccc}
\pi_i(SO(m)) & \longrightarrow & \pi_i(SO(m+k)) & \longrightarrow & \pi_i(V_{k,m}) & \longrightarrow & \pi_{i-1}(SO(m)) \\
\downarrow{\scriptstyle 1} & & \downarrow{\scriptstyle i_*} & & \downarrow{\scriptstyle j_*} & & \downarrow{\scriptstyle 1} \\
\pi_i(SO(m)) & \longrightarrow & \pi_i(SO(m+k+1)) & \longrightarrow & \pi_i(V_{k+1,m}) & \longrightarrow & \pi_{i-1}(SO(m)).
\end{array}$$

By (IV.1.7), i_* is an isomorphism for $i<m+k-1$, and since $k \geqq 2$, it follows that j_* is an isomorphism for $i \leqq m$. □

The following theorem describes what can be accomplished by surgery below the middle dimension. It is closely related to theorems of Mazur [40] and Brown [15]. The proof given here is essentially a translation into the category of differentiable manifolds of an analogous theorem in the category of CW complexes due to Moore [16, Exposé 22 Appendix].

IV.1.13 Theorem. *Let $(M^m, \partial M^m)$ be a smooth compact m-manifold with boundary, $m \geqq 4$, v^k the normal bundle for $(M, \partial M) \subset (D^{m+k}, S^{m+k-1})$, $k >> m$. Let A be a finite complex, $B \subset A$, η^k a k-plane bundle over A, let $f:(M, \partial M) \to (A, B)$ and let $b: v \to \eta$ be a linear bundle map covering f. Then there is a cobordism W of M, with $\partial W = M \cup (\partial M \times I) \cup M'$, $\partial M' = \partial M \times 1$, an extension F of f, $F:(W, \partial M \times I) \to (A, B)$ with $F|\partial M \times t = f|\partial M$ for each $t \in I$, and an extension $\bar{b}$ of b, $\bar{b}: \omega \to \eta$, ω = normal bundle of W in $D^{m+k} \times I$ such that $f' = F|M' : M' \to A$ is $\left[\frac{m}{2}\right]$-connected.*

(We shall call the cobordism of the above type a normal cobordism, in a slight abuse of language.)

(We recall that $[a]$ = greatest integer $\leqq a$, for a real number a.)

Proof. Let us assume by induction that $f: M \to A$ is n-connected, $n+1 \leqq \left[\frac{m}{2}\right]$, and show how to obtain W, F etc. as above, with $f': M' \to A$ $(n+1)$-connected.

If $n+1=0$, we must only show how to make π_0 map onto. Since A is a finite complex, A has only a finite number of components, $A=A_1\cup\cdots\cup A_r$. Let $a_i\in A_i$, and take $M'=M\cup S_1^m\cup\cdots\cup S_r^m$, where S_i^m is an m-sphere. Let $W=M\times I\cup D_1^{m+1}\cup\cdots\cup D_r^{m+1}$ and let $F:W\to A$ be defined by $F|M\times t=f$ for each $t\in I$, $F(D_i^m)=a_i$. Since the normal bundle of D^m is trivial, and the extension condition on the bundle map is easy to fulfill on the D_i^m, it follows easily that b extends to $\bar{b}$ over W. Clearly $f'=F|M'$ is onto $\pi_0(A)$, which proves the first step of our induction.

Now assume $n=1$, $f:M\to A$ is 0-connected. Let M_1 and M_2 be two components of M such that $f(M_1)$ and $f(M_2)$ are in the same component of A. Take two points $x_i\in$ interior M_i, $i=1,2$, and define $\bar{\varphi}:S^0\to M$ by $\bar{\varphi}(1)=x_1$, $\bar{\varphi}(-1)=x_2$. Since $\bar{\varphi}(S^0)\subset$ a single component of A, it follows that $f:M\to A$ extends to $\bar{f}:M\bigcup_{\bar{\varphi}} D^1\to A$. Then since $m\geqq 4$, it follows from (IV.1.6) and (IV.1.12), that $\bar{\varphi}$ extends to $\varphi:S^0\times D^m\to M$ defining a normal cobordism of f to f' and reducing the number of components of M. Using this argument repeatedly, we arrive at a $1-1$ correspondence of components.

Now we consider the fundamental groups. Take presentations, i.e. systems of generators and relations, $\pi_1(A)=\{a_1,\ldots,a_s;r_1,\ldots,r_t\}$, $\pi_1(M)=\{x_1,\ldots,x_k;y_1,\ldots,y_l\}$ so that r_i is a word in $a_1,\ldots,a_s$ (y_i is a word in $x_1,\ldots,x_k$) each i, and $\pi_1(A)$ is the quotient of the free group on $a_1,\ldots,a_s$ by the smallest normal subgroup containing $r_1,\ldots,r_t$, (similarly for $\pi_1(M)$). Now take s disjoint embeddings of S^0 in an m-cell $D^m\subset\operatorname{int}M$, $\varphi':\bigcup_s S^0\to M$ and assume $f(D^m)=*$, the base point of A. We assume the base point of M is in D^m. Consider $\bar{M}=M\bigcup_{\varphi'}\bigcup_s D^1$. Then $\pi_1(\bar{M})=\pi_1(M)*F$ where F is a free group on s generators $g_1,\ldots,g_s$, where each g_i is the homotopy class of a loop in $D^m\cup\bigcup_s D^1$ consisting of a path in D^m, one of the D^1's, and another path in D^m. Hence $\pi_1(\bar{M})=\{x_1,\ldots,x_k,g_1,\ldots,g_s;y_1,\ldots,y_l\}$.

Define $\bar{f}:\bar{M}\to A$ extending f by letting the image of the i-th D^1 traverse a loop representing the generator a_i. Then $\bar{f}_*:\pi_1(\bar{M})\to\pi_1(A)$ is onto, and furthermore we may represent $\bar{f}_*$ on the free groups by a function $\alpha(x_i)=x_i'$, x_i' a word in $a_1,\ldots,a_s$, and $\alpha(g_i)=a_i$. Then as above we may extend φ' to $\varphi:\bigcup_s S^0\times D^m\to M$ to define a normal cobordism of f, and with $W_\varphi\cong\bar{M}$, and $F:W_\varphi\to A$ homotopic to $\bar{f}:\bar{M}\to A$. (Here W_φ is the trace of the simultaneous surgeries.) By (IV.1.2), $\pi_1(M')\cong\pi_1(W_\varphi)$ where $\partial W_\varphi=M\cup(\partial M\times I)\cup M'$, and hence $f_*':\pi_1(M')\to\pi_1(A)$ is onto, $\pi_1(M')$ has the same presentation as $\pi_1(\bar{M})$, and f_*' is also represented by the function α on the free groups. In particular f' is 1-connected.

Let us consider the exact sequence of the map $f: M \to A$ in homotopy,

$$\cdots \to \pi_{n+1}(f) \to \pi_n(M) \to \pi_n(A) \to \pi_n(f) \to \cdots .$$

Recall that the elements of the groups $\pi_{n+1}(f)$ are defined by commutative diagrams

$$\begin{array}{ccc} S^n & \xrightarrow{\alpha} & M \\ {\scriptstyle k}\downarrow & & \downarrow{\scriptstyle f} \\ D^{n+1} & \xrightarrow{\beta} & A \end{array} \qquad (*)$$

where k is inclusion of the boundary and all maps and homotopies are base point preserving (see [28]). Thus β defines a map $\bar{f}: M \bigcup_\alpha D^{n+1} \to A$ extending f.

IV.1.14 Lemma. *Let $f: M \to A$ be n-connected, $n > 0$, and let $(\beta, \alpha) \in \pi_{n+1}(f)$ be the element represented by the above diagram $(*)$. If $\bar{f}: M \bigcup_\alpha D^{n+1} \to A$ is defined by β as above, then $\pi_i(\bar{f}) = \pi_i(f) = 0$ for $i \leqq n$, and $\pi_{n+1}(\bar{f}) = \pi_{n+1}(f)/K$, where K is a normal subgroup containing the $\pi_1(M)$ module generated by the element (β, α) in $\pi_{n+1}(f)$.*

Proof. Consider the commutative diagram

$$\begin{array}{ccccccc} \to \pi_{l+1}(f) & \longrightarrow & \pi_l(M) & \xrightarrow{f_*} & \pi_l(A) \to \\ \downarrow{\scriptstyle j_*} & & \downarrow{\scriptstyle i_*} & & \downarrow{\scriptstyle 1} \\ \to \pi_{l+1}(\bar{f}) & \longrightarrow & \pi_l\left(M \bigcup_\alpha D^{n+1}\right) & \xrightarrow{F_*} & \pi_l(A) \to . \end{array}$$

Here $i: M \to M \bigcup_\alpha D^{n+1}$ is inclusion, and j_* is induced by $(1, i)$ on the diagram $(*)$. Clearly i_* is an isomorphism for $l < n$ and onto for $l = n$, so it follows easily that $\pi_l(\bar{f}) = \pi_l(f) = 0$ for $l \leqq n$.

Clearly any map of S^n into $M \bigcup_\alpha D^{n+1}$ is homotopic to a map into M, so that any pair (β', α')

$$\begin{array}{ccc} S^n & \xrightarrow{\alpha'} & M \bigcup_\alpha D^{n+1} \\ \downarrow & & \downarrow{\scriptstyle \bar{f}} \\ D^{n+1} & \xrightarrow{\beta'} & A \end{array}$$

is homotopic to another $(\beta'', i\alpha'')$

$$\begin{array}{ccccc} S^n & \xrightarrow{\alpha''} & M & \xrightarrow{i} & M \bigcup_\alpha D^{n+1} \\ \downarrow & & \downarrow{\scriptstyle f} & & \downarrow{\scriptstyle \bar{f}} \\ D^{n+1} & \xrightarrow{\beta''} & A & \xrightarrow{1} & A . \end{array}$$

Hence $j_*: \pi_{n+1}(f) \to \pi_{n+1}(\bar{f})$ is onto. Clearly (β, α) is in the kernel j_* and hence everything obtained from (β, α) by the action of $\pi_1(M)$ is also in kernel j_*, which proves the lemma. □

Now we recall that from our previous work we may assume $f: M \to A$ is 1-connected and that furthermore the fundamental groups have presentations $\pi_1(M) = \{x_1, \ldots, x_k, g_1, \ldots, g_s; y_1, \ldots, y_l\}$, y_i words in $x_1, \ldots, x_k$, and $\pi_1(A) = \{a_1, \ldots, a_s; r_1, \ldots, r_t\}$, with $f_*: \pi_1(M) \to \pi_1(A)$ presented by the function $\alpha(x_j) = x_j'(\bar{a})$, a word in $a_1, \ldots, a_s$, $j = 1, \ldots, k$, $\alpha(g_i) = a_i$, $i = 1, \ldots, s$.

IV.1.15 Lemma. *kernel f_* is the smallest normal subgroup containing the words $x_j^{-1}(x_j'(\bar{g}))$, $j = 1, \ldots, k$ and $r_i(\bar{g})$, $i = 1, \ldots, t$, where $x_j'(\bar{g})$ and $r_i(\bar{g})$ are the words in $a_1, \ldots, a_s$ with the a_i's replaced by g_i's.*

Proof. Adding the relations $x_j^{-1}(x_j'(\bar{g}))$ makes $g_1, \ldots, g_s$ into a set of generators. Adding the relations $r_i(\bar{g})$ makes the group into $\pi_1(A)$, with α defining the isomorphism. The map α annihilates $x_j^{-1}(x_j'(\bar{g}))$ and $r_i(\bar{g})$ so that these elements generate kernel f_* as a normal subgroup. □

Now for each element $x_j^{-1}(x_j'(\bar{g}))$ and $r_i(\bar{g})$ choose an element $\bar{x}_j, \bar{r}_i \in \pi_2(f)$ such that $\partial \bar{x}_j = x_j^{-1}(x_j'(\bar{g}))$, $\partial \bar{r}_i = r_i(\bar{g})$, and choose representatives $\bar{x}_j$, $\bar{r}_i$ such that they are disjoint embeddings of S^1 into M, which is possible since $m \geqq 4$. Let $\bar{M} = M \cup \bigcup D^2$, with the D^2's attached by these embeddings. It follows from (IV.1.14) that $\bar{f}_*: \pi_1(\bar{M}) \to \pi_1(A)$ is an isomorphism. Using again (IV.1.6) and (IV.1.12), it follows that there is a normal cobordism W, and map $F: W \to A$ such that $\bar{M} \subset W$ is a deformation retract and $F | \bar{M} = \bar{f}$, so that $F_*: \pi_1(W) \to \pi_1(A)$ is an isomorphism. By (IV.1.2) and (IV.1.3) it follows that if M' is the result of the surgery, then $f'_*: \pi_1(M') \to \pi_1(A)$ is an isomorphism, and hence $\pi_2(A) \to \pi_2(f)$ is onto and therefore $\pi_2(f)$ is abelian.

Now we proceed to the induction step. Suppose $f: M \to A$ is n-connected, $n > 0$, and if $n = 1$ suppose $\pi_1(M) \to \pi_1(A)$ is an isomorphism, so that $\pi_2(f)$ is abelian.

IV.1.16 Lemma. *$\pi_{n+1}(f)$ is a finitely generated module over $\pi_1(M)$.*

Proof. If f is replaced by an inclusion $f_0: M \to A'$, where M and A' are still finite complexes, then $\pi_i(f) \cong \pi_i(A', M)$. Since $\pi_i(A', M) = 0$ for $i \leqq n$, all the cells of dimension $\leqq n$ in A' can be deformed into M to get a new A'' such that $r: (A', M) \to (A'', M)$, $r_*: \pi_i(A', M) \cong \pi_i(A'', M)$, and A'' is a finite complex with all cells of dimension $\leqq n$ in M, $A'' = M \cup \bigcup_{i=1}^{s} D_i^{n+1}$ $\cup$ cells of higher dimension. Let $\tilde{M}$, $\tilde{A}''$ be the universal coverings of M, A''. Then $\pi_i(\tilde{A}'', \tilde{M}) \cong \pi_i(A'', M)$ and since $\tilde{A}''$ and $\tilde{M}$ are 1-connected $\pi_i(\tilde{A}'', \tilde{M}) \cong H_i(\tilde{A}'', \tilde{M})$ as $\pi_1(M)$ modules, by the Relative Hurewicz

Theorem. But clearly the preimages of the $(n+1)$ cells of A'' are the $(n+1)$ cells of $\tilde{A}''$, so that modulo the action of $\pi_1(M)$ there are only a finite number of them. Hence $H_{n+1}(\tilde{A}'', \tilde{M})$ is finitely generated over $\pi_1(M)$ and the lemma follows. □

Now we may represent each of this finite number of elements in $\pi_{n+1}(f)$ by a map

$$\begin{array}{ccc} S^n & \xrightarrow{\alpha_i} & M \\ \downarrow & & \downarrow f \\ D^{n+1} & \xrightarrow{\beta_i} & A. \end{array}$$

If $n+1 \leqq \left[\frac{m}{2}\right]$, then $n < \frac{m}{2}$ and it follows from Whitney's embedding theorem ("general position") that we may choose (β_i, α_i) so that the α_i are disjoint embeddings. Setting $\bar{M} = M \cup \bigcup_i D_i^{n+1}$, D_i^{n+1} attached by α_i, $\bar{f}: \bar{M} \to A$ defined by the β_i's, we may apply (IV.1.6) and (IV.1.12) to thicken $\bar{M}$ into a normal cobordism W of M, and using (IV.1.14), $\pi_l(\bar{f}) = 0$ for $l \leqq n+1$. If M' is the other end of W (the result of the surgeries), from (IV.1.2) and (IV.1.3) it follows that $\pi_i(f') = \pi_i(\bar{f}) = 0$ for $i \leqq n+1$. This completes the proof of (IV.1.13). □

Note that we have always used the low dimensionality of the groups involved to ensure that the obstruction $\mathcal{O}$ was zero (IV.1.12) and to get representatives of elements of $\pi_{n+1}(f)$ which were embeddings. These are two difficulties which must be treated in order to get stronger theorems in higher dimensions.

§ 2. The Fundamental Theorem: Preliminaries

Let (A, B) be an oriented Poincaré pair of dimension m, let M be an oriented compact smooth m-manifold with boundary ∂M, and let $f: (M, \partial M) \to (A, B)$ be a map of degree 1. Let η^k be a linear k plane bundle over A, $k >> m$, and let ν^k be the normal bundle of

$$(M, \partial M) \subset (D^{m+k}, S^{m+k-1}).$$

Suppose $b: \nu \to \eta$ is a linear bundle map lying over f. Recall that in Chapter II we called (f, b) a normal map, and we defined a normal cobordism of (f, b) rel B as a $(m+1)$-manifold W with

$$\partial W = M \cup (\partial M \times I) \cup M',$$

together with an extension of f, $F: (W, \partial M \times I) \to (A, B)$ such that $F | \partial M \times t = f | \partial M$ for each $t \in I$, and an extension $\bar{b}$ of b to the normal bundle ω of W in $D^{m+k} \times I$.

Suppose now that A is a 1-connected CW complex, $m \geqq 5$, and that $(f|\partial M)_*: H_*(\partial M) \to H_*(B)$ is an isomorphism.

IV.2.1 Theorem. *There is a normal cobordism rel B of (f, b) to (f', b') such that $f': M' \to A$ is $\left[\frac{m}{2}\right]+1$ connected if and only if $\sigma(f, b)=0$. In particular if m is odd this is true.*

The proof of this theorem will take up the rest of Chapter IV. First we note the corollary:

IV.2.2 Corollary (Fundamental Theorem). *The map f' above is a homotopy equivalence. Hence (f, b) is normally cobordant rel B to a homotopy equivalence if and only if $\sigma(f, b)=0$. In particular it is true if m is odd.*

Proof of Corollary. Look at the map of exact sequences

$$\begin{array}{ccccccccc}
\longrightarrow & H_i(\partial M') & \longrightarrow & H_i(M') & \longrightarrow & H_i(M', \partial M') & \longrightarrow & H_{i-1}(\partial M') \\
& \downarrow{\scriptstyle (f'|\partial M')_*} & & \downarrow{\scriptstyle f'_*} & & \downarrow{\scriptstyle \bar{f}'_*} & & \downarrow{\scriptstyle (f'|\partial M')_*} \\
\longrightarrow & H_i(B) & \longrightarrow & H_i(A) & \longrightarrow & H_i(A, B) & \longrightarrow & H_{i-1}(B).
\end{array}$$

By hypothesis, $(f|\partial M)_* \, H_*(\partial M) \to H_*(B)$ is an isomorphism, and $\partial M' = \partial M$, $f'|\partial M' = f|\partial M$, so $(f'|\partial M')_*$ is an isomorphism in each dimension i. Since $f': M' \to A$ is $\left[\frac{m}{2}\right]+1$-connected, $f'_*: H_i(M') \to H_i(A)$ is an isomorphism for $i \leqq \frac{m}{2}$. Hence by the Five Lemma,

$$\bar{f}'_*: H_i(M', \partial M') \to H_i(A, B)$$

is an isomorphism for $i \leqq \frac{m}{2}$. Since f' is a map of degree 1, it follows from Poincaré duality that $f'^*: H^j(A) \to H^j(M')$ is an isomorphism for $j \geqq m - \frac{m}{2} = \frac{m}{2}$ (see (I.2.6)). Now $f'^*: H^j(A) \to H^j(M')$ is given, by the Universal Coefficient Theorem, by $f'^* = \operatorname{Hom}(f'_{*j}, \mathbb{Z}) + \operatorname{Ext}(f'_{*j-1}, \mathbb{Z})$, where $f'_{*i}: H_i(M') \to H_i(A)$. Since f'_{*i} is an isomorphism for $i \leqq \frac{m}{2}$ it follows that $f'^*: H^j(A) \to H^j(M')$ is an isomorphism for $j \leqq \frac{m}{2}$, and hence $f'^*: H^j(A) \to H^j(M')$ is an isomorphism for all j. Hence $H^*(f')=0$, so by the Universal Coefficient Theorem $H_*(f')=0$, and since M' and A are 1-connected, by the Relative Hurewicz Theorem and the Theorem of Whitehead, $f': M' \to A$ is a homotopy equivalence.

The remainder of § 2 will be devoted to the preliminaries of the proof of (IV.2.1).

By (IV.1.13), we may assume that $f: M \to A$ is $\left[\frac{m}{2}\right]$-connected, i.e. $\pi_i(f)=0$ for $i \leqq \left[\frac{m}{2}\right]$. Set $l=\left[\frac{m}{2}\right]$. Since A, M are 1-connected, it follows from the Relative Hurewicz Theorem that $\pi_{l+1}(f) \cong H_{l+1}(f)$. Then we have a commutative diagram:

$$\begin{array}{ccccccc}
\longrightarrow \pi_{l+1}(f) & \longrightarrow & \pi_l(M) & \xrightarrow{f_\#} & \pi_l(A) & \longrightarrow & 0 \\
\downarrow h \cong & & \downarrow h & & \downarrow h & & \\
\longrightarrow H_{l+1}(f) & \longrightarrow & H_l(M) & \xrightarrow{f_*} & H_l(A) & \longrightarrow & 0
\end{array}$$

where h is the Hurewicz homomorphism, and we use $f_\#$ to denote the map of homotopy groups induced by f. We recall that f_* is onto and splits by (I.2.5). It follows that $(\text{kernel}\, f_*)_l = h(\text{kernel}\, f_\#)_l$.

We recall Whitney's embedding theorem (see [42] for a proof): *Let $c: V^n \to M^m$ be a continuous map of smooth manifolds, $m \geqq 2n$, $m-n>2$, M 1-connected, V connected. Then c is homotopic to a smooth embedding.*

Since $l \leqq \frac{1}{2}m$, it follows from Whitney's embedding theorem that any element $x \in \pi_{l+1}(f)$ may be represented by $(\beta, \bar{\varphi})$, where $\bar{\varphi}$ is a smooth embedding of S^l in Interior M, and $\beta: D^{l+1} \to A$, $\beta i = f\bar{\varphi}$. Let $\bar{M} = M \bigcup_{\bar{\varphi}} D^{l+1}$, $\bar{f}: \bar{M} \to A$ extending f, defined by β.

Now we have two problems to consider:

(1) If $m=2l$, then the obstruction $\mathcal{O}$ to thickening $\bar{M}$, $\bar{f}$ to a normal cobordism lies in a non-zero group $\pi_l(V_{k,l})$ (see (IV.1.6) and (IV.1.12)).

(2) Though (IV.1.14) tells us how to compute $\pi_{l+1}(\bar{f})$, the relation between this and $\pi_{l+1}(f')$ is no longer obvious if $l=\left[\frac{m}{2}\right]$, where f' is the map on the result of the surgery (c.f. (IV.1.2)).

The remainder of § 2 will be devoted to some preliminary results on question (2).

For the remainder of this paragraph we assume (f, b) is a normal map satisfying the hypotheses of (IV.2.1) and $f: M \to A$ is q-connected where $\left[\frac{m}{2}\right] = q$, so that $m = 2q$ or $2q+1$.

IV.2.3 Lemma. *f is $(q+1)$ connected if and only if $f_*: H_q(M) \to H_q(A)$ is an isomorphism, i.e. if $K_q(M)=0$.*

Proof. By the Relative Hurewicz Theorem $\pi_{q+1}(f) \cong H_{q+1}(f)$, and by (I.2.5), $f_*: H_{q+1}(M) \to H_{q+1}(A)$ is onto so that

$$H_{q+1}(f) \cong (\ker f_*)_q = K_q(M). \quad \square$$

Thus we shall study the effect of surgery on homology. To simplify our arguments we will use the following lemma, which reduces the problem to the case of closed manifolds.

Let (f_1, b_1), (f_2, b_2) be two disjoint copies of the normal map (f, b), so that $f_i: (M_i, \partial M_i) \to (A_i, B_i)$ is f renamed, etc., $i = 1, 2$. Then by (I.3.2), $A_3 = A_1 \cup A_2$ with B_1 identified to B_2 is a Poincaré complex (the "double" of A) $M_3 = M_1 \cup M_2$ along $\partial M_1 = \partial M_2$ is a smooth closed oriented manifold, and $f_3 = f_1 \cup f_2$, $b_3 = b_1 \cup b_2$ defines a normal map (f_3, b_3), $f_3: M_3 \to A_3$. Further it is easy to see from the Mayer-Vietoris sequences (since $(f \mid \partial M)_*$ is an isomorphism) that

IV.2.4 $H_i(f_3) = 0$ for $i < l+1$ and

$$H_{q+1}(f_3) \cong K_q(M_3) \cong K_q(M_1) + K_q(M_2).$$

Now suppose $\varphi: S^q \times D^{m-q} \to \operatorname{int} M_1$ is a smooth embedding such that $f_1 \circ \varphi \sim *$ and such that φ defines a normal surgery on M_1 and by inclusion on M_3 (with respect to (f_1, b_1) and (f_3, b_3)). If a prime denotes the result of surgery then we have

IV.2.5 $M_3' = M_1' \cup M_2$ and $K_q(M_3') \cong K_q(M_1') + K_q(M_2)$.

This follows easily from the fact that we have not changed the factor M_2 in the decomposition of M_3.

Hence we get:

IV.2.6 Proposition. *The effect of a normal surgery on $K_q(M)$ is the same as the effect of the induced surgery on $K_q(M_3)$, and hence to compute its effect we may assume $\partial M = B = \emptyset$.*

This will simplify the algebra in our discussion.

Let $\varphi: S^q \times D^{m-q} \to \operatorname{int} M$ be a smooth embedding which defines a normal surgery on M (with respect to (f, b)). Set $M_0 = M - \operatorname{int} \varphi(S^q \times D^{m-q})$, and let $M' = M_0 \cup D^{q+1} \times S^{m-q-1}$, with $\varphi(S^q \times S^{m-q-1})$ identified with $S^q \times S^{m-q-1} = \partial(D^{q+1} \times S^{m-q-1})$. Then M' is the result of the surgery on M. Since φ defines a normal surgery, $H_q(M') = H_q(A) + K_q(M')$, and we wish to calculate the change of $K_q(M)$ to $K_q(M')$ which is the same as the change of $H_q(M)$ *to* $H_q(M')$.

Now we recall some useful facts relating Poincaré duality in manifolds and submanifolds.

IV.2.7 Proposition. *Let U be a compact m-manifold with boundary, $f: U \subset \operatorname{int} W$, W a compact m-manifold with boundary,*

$$g: (W, \partial W) \subset (W, W - \operatorname{int} U),$$

oriented compatibly. Then the diagram below is commutative:

$$\begin{array}{ccccc}
H^q(W,\partial W) & \xleftarrow{g^*} & H^q(W, W-\operatorname{int} U) & \xrightarrow{\cong} & H^q(U,\partial U) \\
\downarrow {\scriptstyle [W]\cap} & & \downarrow {\scriptstyle (g_*[W])\cap} & & \downarrow {\scriptstyle [U]\cap} \\
H_{m-q}(W) & \xrightarrow{1} & H_{m-q}(W) & \xleftarrow{f_*} & H_{m-q}(U)
\end{array}$$

so if $x \in H^q(U/\partial U)$, $f_*([U]\cap x) = [W]\cap \bar{g}^*(x)$, *where* $\bar{g}: W/\partial W \to U/\partial U$, *(interpreting the cap products appropriately).*

Proof. If $\bar{f}:(U,\partial U)\to(W, W-\operatorname{int} U)$, then $\bar{f}_*[U] = g_*[W]$, since we have oriented U and W compatibly. Then the commutativity follows from the naturality of cap product (see Chapter I, § 1). □

IV.2.8 Corollary. *Let* E = *normal tube of* $f: N^n \subset W^m$, N *closed oriented and let* $\bar{g}: W/\partial W \to E/\partial E = T(\nu)$, ν = *normal bundle of* $N^n \subset W^m$. *Let* $U \in H^{m-n}(T(\nu))$ *be the Thom class. Then*

$$[W]\cap \bar{g}^* U = f_*[N].$$

Proof. Since $[E]\cap U = [N]$, by (IV.2.7),

$$f_*([E]\cap U) = f_*[N] = [W]\cap(\bar{g}^*(U)). \quad \square$$

Recall now the definition of the intersection pairing in homology:

$$\cdot : H_q(M)\otimes H_{m-q}(M,\partial M)\to \mathbb{Z}$$

defined by $x\cdot y = (x', y') = (x'\cup y')[M]$ where $x' \in H^{m-q}(M,\partial M)$, $y' \in H^q(M)$ such that $[M]\cap x' = x \in H_q(M)$, $[M]\cap y' = y \in H_{m-q}(M,\partial M)$. This induces an intersection product

$$\cdot : H_q(M)\otimes H_{m-q}(M)\to \mathbb{Z}$$

by

$$x\cdot y = x\cdot j_*(y),\ j: M\to (M,\partial M).$$

The properties of the pairing (,) on cohomology induce analogous properties for the intersection pairing, such as

(a) With coefficients in a field F, $H_q(M;F)\otimes H_{m-q}(M,\partial M;F)\to F$ is a non-singular pairing. (This also holds over $\mathbb{Z}$, modulo torsion.)

(b) If $x\in H_q(M)$, $y \in H_{m-q}(M)$, $x\cdot y = (-1)^{q(m-q)} y\cdot x$.

IV.2.9 Proposition. *Let* $x\in H_q(M)$, $y\in H_{m-q}(M,\partial M)$, $x'\in H^{m-q}(M,\partial M)$, $y'\in H^q(M)$ *such that* $[M]\cap x' = x$, $[M]\cap y' = y$. *Then* $x\cdot y = x'(y)$, *(i.e. evaluation of the cohomology class* x' *on the homology class* y*).*

Proof. $x\cdot y = (x'\cup y')[M] = x'([M]\cap y') = x'(y)$, using (I.1.1). □

Now let $\varphi: S^q \times D^{m-q} \to \operatorname{int} M$ be a smooth embedding. Set $E = S^q \times D^{m-q}$, $M_0 = M - \varphi \operatorname{int} E$, $M' = M_0 \cup (D^{q+1} \times S^{m-q-1})$ the result of the surgery based on φ.

Following [34] we will consider the exact sequences of the pairs (M, M_0) and (M', M_0).

As usual we have the excision $\varphi: (E, \partial E) \to (M, M_0)$ which induces isomorphisms on the relative homology and cohomology groups. Thinking of E as the normal tube of $S^q \subset M$, let $U \in H^{m-q}(E, \partial E) = \mathbb{Z}$ be the Thom class, a generator. If $\mu = [E] \cap U$, then $\mu = i_*[S^q]$, $i: S^q \subset E$, and $\mu \cdot x = U(x)$, $x \in H_{m-q}(E, \partial E)$, by (IV.2.9), induces an isomorphism $H_{m-q}(E, \partial E) \to \mathbb{Z}$ by property (a) above. Let $j: M \to (M, M_0)$ be the inclusion.

IV.2.10 Proposition. $\mu \cdot (j_*(y)) = (\varphi_*(\mu)) \cdot y$.

Proof. $\mu \cdot (j_*(y)) = U(j_*(y)) = (j^* U)(y) = (\varphi_*(\mu)) \cdot y$, using (IV.2.9) and (IV.2.8), and identifying $j_*: H_*(M) \to H_*(M, M_0)$ with the collapsing map $\bar{j}_*: H_*(M) \to H_*(M/M_0) = H_*(E/\partial E)$. □

IV.2.11 Corollary. *The following sequence is exact:*

$$\cdots 0 \to H_{m-q}(M_0) \to H_{m-q}(M) \xrightarrow{x\cdot} \mathbb{Z} \xrightarrow{d} H_{m-q-1}(M_0) \to H_{m-q-1}(M) \to 0$$

where $x = \varphi_*(\mu)$, $\mu \in H_q(S^q \times D^{m-q})$ *is the image of* $[S^q]$ *the orientation class of* S^q.

Proof. The sequence is that of (M, M_0), replacing $H_{m-q}(M, M_0)$ by $\mathbb{Z}$ using the diagram

$$\begin{array}{ccc} H_{m-q}(E, \partial E) & \xrightarrow{\cong} & H_{m-q}(M, M_0) \\ \downarrow{\scriptstyle \mu\cdot} & & \\ \mathbb{Z} & & \end{array}$$

and using (IV.2.10) to identify $x\cdot$. □

Thus there is also an exact sequence

$$0 \to H_{q+1}(M_0) \to H_{q+1}(M') \xrightarrow{y\cdot} \mathbb{Z} \xrightarrow{d'} H_q(M_0) \xrightarrow{i'_*} H_q(M') \to 0$$

where $y = \psi_*(\mu')$, $\mu' = k'_*[S^{m-q-1}]$ generates $H_{m-q-1}(D^{q+1} \times S^{m-q-1})$, $\psi: D^{q+1} \times S^{m-q-1} \to M'$ is the natural embedding,

$$k': S^{m-q-1} \to D^{q+1} \times S^{m-q-1}$$

is inclusion.

Let λ be the generator of $\mathbb{Z}$ above corresponding to

$$\lambda \in H_{r+1}(S^q \times D^{r+1}, S^q \times S^r) = \mathbb{Z},$$

such that $U(\lambda) = 1$, (similarly for λ').

IV.2.12 Lemma. *$i_* d'(\lambda') = \varphi_*(\mu) = x$, and $i'_* d(\lambda) = \psi_*(\mu') = y$.*

Proof. Let $m = q + r + 1$. We have the commutative diagram

$$\begin{array}{ccccc}
\to H_{r+1}(S^q \times D^{r+1}, S^q \times S^r) & \xrightarrow{\partial_1} & H_r(S^q \times S^r) & \xrightarrow{i_{1*}} & H_r(S^q \times D^{r+1}) \to \\
\downarrow{\scriptstyle \varphi_*} & & \downarrow{\scriptstyle \varphi_{0*}} & & \downarrow{\scriptstyle \varphi_*} \\
\to H_{r+1}(M, M_0) & \xrightarrow{\partial} & H_r(M_0) & \xrightarrow{i_*} & H_r(M).
\end{array}$$

Clearly if $\lambda \in H_{r+1}(S^q \times D^{r+1}, S^q \times S^r)$ such that $U(\lambda) = 1$, then $\partial_1 \lambda = 1 \otimes [S^r] \in H_r(S^q \times S^r)$. We also have the commutative diagram

$$\begin{array}{ccc}
H_*(S^q \times S^r) & \xrightarrow{i_{2*}} & H_*(D^{q+1} \times S^r) \\
\downarrow{\scriptstyle \varphi_{0*}} & & \downarrow{\scriptstyle \psi_*} \\
H_*(M_0) & \xrightarrow{i'_*} & H_*(M')
\end{array}$$

and $i_{2*}(1 \otimes [S^r]) = \mu'$. Hence

$$\begin{aligned}
i'_* d(\lambda) &= i'_* \partial \varphi_*(\lambda) = i'_* \varphi_{0*} \partial_1(\lambda) \\
&= \psi_* i_{2*}(1 \otimes [S^r]) = \psi_*(\mu') = y.
\end{aligned}$$

A similar argument proves the other assertion. □

IV.2.13 Theorem. *Let $\varphi: S^q \times D^{r+1} \to M$ be an embedding, M^m closed, $m = q + r + 1$, $q \leqq r + 1$. Suppose $\bar{\varphi}_*([S^q]) = \varphi_*(\mu) = x$ generates an infinite cyclic direct summand of $H_q(M)$. Then rank $H_q(M') <$ rank $H_q(M)$ and torsion $H_q(M') \cong$ torsion $H_q(M)$, i.e. the free part of $H_q(M)$ is reduced, and the torsion part of $H_q(M)$ is not increased. Further $H_i(M') \cong H_i(M)$ for $i < q$.*

IV.2.14 Corollary. *Let (f, b) be a normal map, $f: (M, \partial M) \to (A, B)$, $(f \mid \partial M)_*$ an isomorphism, and let $\varphi: S^q \times D^{r+1} \to \operatorname{int} M$ be an embedding which defines a normal cobordism of (f, b), $q \leqq r + 1$. Suppose $\varphi_*(\mu) = x$ generates an infinite cyclic direct summand of $K_q(M)$. Then*

$$\text{rank}\, K_q(M') < \text{rank}\, K_q(M)$$

and torsion $K_q(M') \cong$ torsion $K_q(M)$, while $K_i(M') = K_i(M)$, for $i < q$.

Proof. This follows immediately from (IV.2.13) and (IV.2.6). □

With a field of coefficients F we have analogous results:

IV.2.15 Theorem. *Let* φ, M *be as in* (IV.2.13) *and suppose* $\varphi_*(\mu)=x\neq 0$ *in* $H_q(M;F)$. *Then* $\operatorname{rank}_F H_q(M';F)<\operatorname{rank}_F H_q(M;F)$, *and*

$$H_i(M';F)\cong H_i(M;F) \quad \textit{for} \quad i<q\,.$$

IV.2.16 Corollary. *With hypotheses of* (IV.2.14), *suppose only that* $\varphi_*(\mu)=x\neq 0$ *in* $K_q(M;F)$. *Then* $\operatorname{rank}_F K_q(M';F)<\operatorname{rank}_F K_q(M;F)$ *and* $K_i(M';F)\cong K_i(M;F)$ *for* $i<q$.

The proof of the corollary is similar.

Proof of (IV.2.13). Consider the exact sequence of (IV.2.11):

$$0\to H_{r+1}(M_0)\xrightarrow{i_*} H_{r+1}(M)\xrightarrow{x\cdot}\mathbb{Z}\xrightarrow{d} H_r(M_0)\to H_r(M)\to 0\,.$$

Since x generates an infinite cyclic direct summand, it follows from property (a) of the intersection pairing that there is an element $y\in H_{r+1}(M)$ such that $x\cdot y=1$ (since $\partial M=\emptyset$). Hence $x\cdot$ is onto and we get

IV.2.17

$$i_*: H_r(M_0)\cong H_r(M)$$

$$0\to H_{r+1}(M_0)\xrightarrow{i_*} H_{r+1}(M)\to\mathbb{Z}\to 0\,.$$

Consider the sequence of (IV.2.11) for (M', M_0) and the commutative diagram from (IV.2.12):

IV.2.18

$$\begin{array}{ccccccccccc} 0\to H_{q+1}(M_0)\to H_{q+1}(M')\xrightarrow{y\cdot}\mathbb{Z}\xrightarrow{d'} & H_q(M_0) & \xrightarrow{i'_*} H_q(M')\to 0 \\ \searrow & \downarrow{\scriptstyle i_*} & \\ & H_q(M) & \end{array}$$

where $i_*d'(\lambda')=x$. Since x generates an infinite cyclic direct summand, it follows that i_*d' splits, so that d' splits, and

IV.2.19

$$H_q(M_0)\cong\mathbb{Z}+H_q(M')$$

$$i'_*: H_{q+1}(M_0)\cong H_{q+1}(M')\,.$$

From (IV.2.19), it follows that $\operatorname{rank} H_q(M')=\operatorname{rank} H_q(M_0)-1$, and since $q=r$ or $r+1$, from (IV.2.17) it follows that $\operatorname{rank} H_q(M)\geqq\operatorname{rank} H_q(M_0)$, so that $\operatorname{rank} H_q(M')<\operatorname{rank} H_q(M)$, (the difference being 1 if $q=r$, 2 if $q=r+1$).

From (IV.2.17) it follows that torsion $H_q(M_0)$ is isomorphic to torsion $H_q(M)$, and from (IV.2.19) it follows that

$$\operatorname{torsion} H_q(M_0) \cong \operatorname{torsion} H_q(M').$$

Hence torsion $H_q(M') \cong$ torsion $H_q(M)$. □

The proof of (IV.2.15) is almost identical, using (IV.2.17), (IV.2.18), (IV.2.19) with coefficients in F, and using property (a) of intersection with coefficients in F. We omit the details.

This is as far as one can proceed in the proof of the Fundamental Theorem without considering different dimensions separately, according to parity, or modulo 4. This we shall do in the next sections.

§ 3. Proof of the Fundamental Theorem for m odd

First we note an easy consequence of (IV.2.14).

IV.3.1 Theorem. *Let (f, b) be a normal map, $f:(M, \partial M) \to (A, B)$, A 1-connected, $(f|\partial M)_*: H_*(\partial M) \to H_*(B)$ an isomorphism, $m = 2q+1 \geqq 5$. There is a normal cobordism* rel B *of (f, b) to (f', b'), such that $f': M' \to A$ is q-connected and $K_q(M') \cong$ torsion $K_q(M)$.*

Proof. By (IV.1.13), we may first find a normal cobordism rel B to (f_1, b_1), such that $f_1: M_1 \to A$ is q-connected. We note that the surgeries involved in (IV.1.13) are on spheres of dimension $< q$, so that it follows from (IV.1.2) and (IV.1.3) that $K_q(M_1) \cong K_q(M) + F$, where F is free abelian, and arises from killing torsion classes in K_{q-1}. So let us assume $f: M \to A$ is already q-connected.

Let $x \in K_q(M)$ be a generator of an infinite cyclic direct summand. Since f is q-connected it follows from the Relative Hurewicz Theorem that $\pi_{q+1}(f) \cong H_{q+1}(f)$ and $H_{q+1}(f) \cong K_q(M)$ by (I.2.5). Since $q < \frac{1}{2}m$, it follows from Whitney's embedding theorem that we may represent $x' \in \pi_{q+1}(f)$ by (β, α)

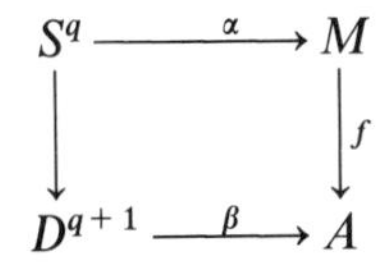

such that α is a smooth embedding. Then β defines a map $\bar{f}: \bar{M} \to A$ where $\bar{M} = M \bigcup_\alpha D^{q+1}$ and by (IV.1.12) since $q < m - q$, the obstruction $\mathcal{O}$ to thickening $\bar{M}$ to a normal cobordism is 0 (see (IV.1.6)). If $x' \in \pi_{q+1}(f)$ is such that α represents $x \in K_q(M)$ then by (IV.2.14), $K_q(M')$ has rank one less than $K_q(M)$ with the same torsion subgroup. Proceeding in this way till the rank is zero, the theorem is proved. □

Now let us put together the two sequences of (IV.2.11), to get the following lemma from [34]:

IV.3.2 Lemma. *We have a diagram*

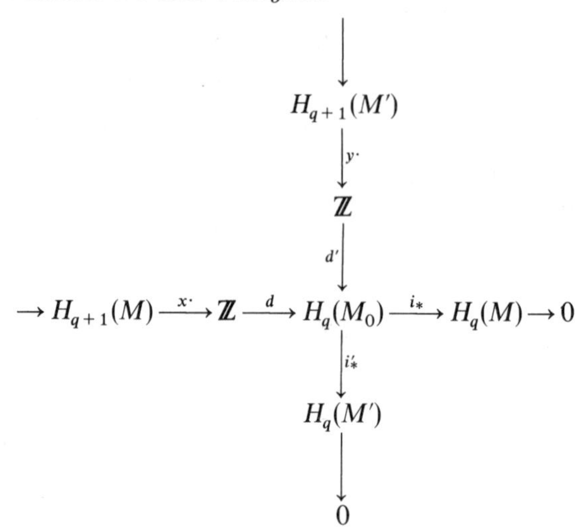

where $i_* d'(\lambda') = x = \varphi_*(\mu)$, $i'_* d(\lambda) = y = \psi_*(\mu')$, μ *a generator of* $H_q(S^q \times D^{q+1})$, μ' *of* $H_q(D^{q+1} \times S^q)$, *etc. Hence*

$$H_q(M')/i'_* d\mathbb{Z} = H_q(M)/i_* d'\mathbb{Z}.$$

Proof. This follows immediately from (IV.2.11), (IV.2.12) and the fact that

$$H_q(M)/i_* d'\mathbb{Z} \cong H_q(M_0)/d'\mathbb{Z} + d\mathbb{Z} \cong H_q(M')/i'_* d(\mathbb{Z}). \quad \square$$

If $x = i_* d'(\lambda')$ has finite order s, then $x\cdot$ is the zero map, so in (IV.3.2)

IV.3.3

$$0 \to \mathbb{Z} \xrightarrow{d} H_q(M_0) \xrightarrow{i_*} H_q(M) \to 0$$

is exact. Also $sd'(\lambda') \in \ker i_* = \operatorname{im} d$, so we have

IV.3.4

$$sd'(\lambda') + td(\lambda) = 0 \quad \text{in} \quad H_q(M_0), \quad \text{some} \quad t \in \mathbb{Z}.$$

IV.3.5 Lemma. *Suppose x is of finite order s in $H_q(M)$. Then y is of infinite order if $t = 0$, and order $y = t$ if $t \neq 0$.*

Proof. Since $d(\lambda)$ is of infinite order by (IV.3.3), (IV.3.4) implies that $d'(\lambda')$ is also of infinite order if $t \neq 0$, (since $s \neq 0$). Clearly

$$ty = ti'_* d(\lambda) = i'_*(-s\, d'(\lambda')) = 0,$$

since $i'_* d' = 0$, and using (IV.3.4). Hence (order y)$|t$.

If $t'y=0$, then $t'i'_* d(\lambda)=i'_*(t'd(\lambda))=0$ so $t'd(\lambda)\in\ker i'_*=\operatorname{im} d'$, and $t'd(\lambda)=-s'd'(\lambda')$ some $s'\in\mathbb{Z}$, or $s'd'(\lambda')+t'd(\lambda)=0$ in $H_q(M_0)$. Applying i_*, we get $s'i_* d'(\lambda')=s'x=0$, so $s'=ls$. Subtracting l times (IV.3.4) from $s'd'(\lambda')+t'd(\lambda)=0$ we get $(t'-lt)d(\lambda)=0$. But $d(\lambda)$ is of infinite order, so $t'-lt=0$ or $t'=lt$. Hence $t\,|\,t'$, and $t=$ order y.

Suppose $t=0$ so that $sd'(\lambda')=0$. Then $\ker i'_*\subset \operatorname{torsion} H_q(M_0)$, so i'_* is $1-1$ on $d\mathbb{Z}$, and hence $y=i'_* d(\lambda)$ is of infinite order in $H_q(M')$. □

Consider the commutative diagram

IV.3.6

$$\begin{array}{ccc} H_q(S^q\times S^q) & \xleftarrow{\ \partial'\ } & H_{q+1}(D^{q+1}\times S^q, S^q\times S^q)=\mathbb{Z} \\ \uparrow{\scriptstyle\partial} & \searrow{\scriptstyle\varphi_{0*}} & \downarrow{\scriptstyle d'} \\ \mathbb{Z}=H_{q+1}(S^q\times D^{q+1}, S^q\times S^q) & \xrightarrow{\ \ d\ \ } & H_q(M_0) \end{array}$$

where d and d' are from the sequences of (IV.2.11). Recall that

$$\lambda\in H_{q+1}(S^q\times D^{q+1}, S^q\times S^q)\quad\text{such that}\quad \partial\lambda=1\otimes[S^q],$$
$$\lambda'\in H_{q+1}(D^{q+1}\times S^q, S^q\times S^q)\quad\text{such that}\quad \partial'\lambda'=[S^q]\otimes 1.$$

Suppose M is closed, so that $\partial M_0=S^q\times S^q$, and $\varphi_0: S^q\times S^q\to M_0$ is the inclusion of the boundary. Then we have the exact sequence diagram of Poincaré duality (I.2.2):

$$\begin{array}{ccccccc} \to H^q(M_0) & \xrightarrow{\varphi_0^*} & H^q(S^q\times S^q) & \xrightarrow{\delta} & H^{q+1}(M_0, S^q\times S^q) \to \\ \downarrow{\scriptstyle [M_0]\cap} & & \downarrow{\scriptstyle [S^q\times S^q]\cap} & & \downarrow{\scriptstyle [M_0]\cap} \\ \to H_{q+1}(M_0, S^q\times S^q) & \xrightarrow{\partial_0} & H_q(S^q\times S^q) & \xrightarrow{\varphi_{0*}} & H_q(M_0) \to \end{array}$$

Thus

IV.3.7

$$[S^q\times S^q]\cap(\text{image }\varphi_0^*)=\text{kernel }\varphi_{0*}.$$

By (IV.3.6), $d'(\lambda')=\varphi_{0*}\,\partial'(\lambda')=\varphi_{0*}([S^q]\otimes 1)$, and

$$d(\lambda)=\varphi_{0*}\,\partial(\lambda)=\varphi_{0*}(1\otimes[S^q]),$$

so that

$$\varphi_{0*}(s([S^q]\otimes 1)+t(1\otimes[S^q]))=0.$$

IV.3.8 Lemma. *Let q be even. Then $\varphi_{0*}(s([S^q]\otimes 1)+t(1\otimes[S^q]))=0$ implies either $s=0$ or $t=0$.*

Proof. Let $U\in H^q(S^q)$ such that $U[S^q]=1$. Then

$$[S^q\times S^q]\cap(U\otimes 1)=1\otimes[S^q]\quad\text{and}\quad [S^q\times S^q]\cap(1\otimes U)=[S^q]\otimes 1$$

in $H_q(S^q \times S^q)$. Hence

$$[S^q \times S^q] \cap (s(1 \otimes U) + t(U \otimes 1)) = s([S^q] \otimes 1) + t(1 \otimes [S^q]),$$

and by (IV.3.7) it follows that $s(1 \otimes U) + t(U \otimes 1) = \varphi_0^*(z)$, for some $z \in H^q(M_0)$. But $\varphi_0^* : H^{2q}(M_0) \to H^{2q}(S^q \times S^q)$ is zero, as φ_0 is the inclusion of the (connected) boundary of M_0. Hence

$$(s(1 \otimes U) + t(U \otimes 1))^2 = \varphi_0^*(z^2) = 0$$

But $(s(1 \otimes U) + t(U \otimes 1))^2 = 2st(U \otimes U)$ if $q = \dim U$ is even. Hence it is zero if and only if either $s = 0$ or $t = 0$. □

Proof of Theorem (IV.2.1) *for* $m = 2q+1$, *q even:*

By (IV.3.1), we may assume $f: M \to A$ is q-connected and $K_q(M)$ is a torsion group. Let $x \in K_q(M)$ generate a cyclic summand of order s. Let $\varphi: S^q \times D^{q+1} \to M$ be an embedding with $\varphi_*(\mu) = x$, and defining a normal cobordism of (f, b). Assume M is closed, using (IV.2.6). Consider diagram (IV.3.2). By (IV.2.12) $i_* d'(\lambda') = x$, a generator of a summand $\mathbb{Z}_s \subset H_q(M)$. By (IV.3.4) and (IV.3.8), $s d'(\lambda') = 0$, so $d'(\lambda')$ generates a cyclic direct summand $\mathbb{Z}_s \subset H_q(M_0)$.

From (IV.3.3) it follows that torsion $H_q(M_0)$ is isomorphic to a subgroup of torsion $H_q(M)$, and since $H_q(M') \cong H_q(M_0)/d'(\mathbb{Z})$, it follows that torsion $H_q(M')$ is isomorphic to a subgroup of torsion $H_q(M)$ with at least one cyclic summand $\mathbb{Z}_s$ missing, so the same is true for $K_q(M')$. (It follows also that $\operatorname{rank} H_q(M') = \operatorname{rank} H_q(M) + 1$.) By (IV.3.1) we may find a normal cobordism of (f', b') to (f'', b'') with $K_q(M'') = \operatorname{torsion} K_q(M') < K_q(M)$. Iterating the above steps, since $K_q(M)$ is finitely generated, eventually this process must terminate, and we get an (f_1, b_1) with $K_q(M_1) = 0$ and f_1 $(q+1)$-connected. □

From now on then, we will assume q is odd.

Let $\varphi: S^q \times D^{q+1} \to M$ be an embedding which defines a normal cobordism, i.e. so that (f, b) extend over the trace W_φ of the surgery based on φ. Let $\omega: S^q \to SO(q+1)$, let $SO(q+1)$ act on the right on D^{q+1}, and define a new embedding $\varphi_\omega: S^q \times D^{q+1} \to M$ by $\varphi_\omega(x, t) = \varphi(x, t\omega(x))$, $x \in S^q$, $t \in D^{q+1}$. Then φ_ω defines a surgery, and the result $M'_\omega = M_0 \cup D^{q+1} \times S^q$, using the diffeomorphism $\omega': S^q \times S^q \to S^q \times S^q$, $\omega'(x, y) = (x, y\omega(x))$, to identify the boundaries.

IV.3.9 Lemma. *The trace of the surgery based on* φ_ω *also defines a normal cobordism if and only if the homotopy class* $\{\omega\}$ *of* ω *goes to zero in* $\pi_q(SO(q+k+1))$, *i.e.* $i_*\{\omega\} = 0$ *where* $i: SO(q+1) \to SO(q+k+1)$ *is inclusion.*

Proof. The map $\varphi_{i\omega}: S^q \times D^{q+1} \times R^k \to M \times R^k$,

$$\varphi_{i\omega}(x, t, r) = (\varphi(x, t\omega(x)), r) = (\varphi_\omega(x, t), r)$$

defines a new framing of the normal bundle to S^q in D^{m+k}, i.e. of $\nu|S^q+\nu'$ where $\nu=$ normal bundle of $M\subset D^{m+k}$, $\nu'=$ normal bundle of $S^q\subset M$. Then φ_ω defines a normal cobordism if and only if this framing extends to a framing of the normal bundle of D^{q+1} in $D^{m+k}\times I$, so that the first part of the frame defines an embedding $D^{q+1}\times D^{q+1}$ in $D^{m+k}\times I$ extending $\varphi_\omega: S^q\times D^{q+1}\subset M\subset D^{m+k}$, and the second part of the frame extends the trivialization of $\nu|\varphi(S^q\times D^{q+1})$ defined by $b:\nu\to\eta$, to a trivialization of the normal bundle of $D^{q+1}\times D^{q+1}$, and hence that of $M\times I\cup D^{q+1}\times D^{q+1}$.

Now $S^q=\partial D^{q+1}$, $D^{q+1}\subset D^{m+k}\times I$ such that the normal bundle of S^q in $D^{m+k}\times 0$ is the restriction to S^q of γ = the normal bundle of $D^{q+1}\subset D^{m+k}\times I$. Now γ has a framing defined on S^q by the map $\hat\varphi: S^q\times D^{q+1}\times R^k\to E(\nu)$, $\hat\varphi(x,t,r)=(\varphi(x,t),r)$, since φ defined a normal cobordism. The difference of these two framings is a map of S^q into $SO(q+k+1)$ which is obviously $i\omega$.

Hence the frame $\varphi_{i\omega}$ extends over D^{q+1} if and only if $i\omega$ is homotopic to zero in $SO(q+k+1)$. □

By (IV.1.7), $\pi_q(SO(q+r))\to\pi_q(SO(q+r+1))$ is an isomorphism for $r>1$, so that $\ker i_*$, $i_*:\pi_q(SO(q+1))\to\pi_q(SO(q+k+1))$ is the same for all $k\geqq 1$. For $k=1$, the exact homotopy sequence of the fibre space $SO(q+1)\xrightarrow{i} SO(q+2)\to S^{q+1}$ gives that $(\ker i_*)_q=\partial_0\,\pi_{q+1}(S^{q+1})$, $\partial_0:\pi_{q+1}(S^{q+1})\to\pi_q(SO(q+1))$ the boundary in the exact sequence. Hence from (IV.3.9) if $\varphi: S^q\times D^{q+1}\to M$ defines a normal cobordism, then we may change φ by $\omega: S^q\to SO(q+1)$ if $\{\omega\}\in\partial_0\pi_{q+1}(S^{q+1})$ and φ_ω will still define a normal cobordism.

Now we will compare the effect of surgeries based on φ and φ_ω. Let $g_1'=[S^q]\otimes 1$, $g_2'=1\otimes[S^q]\in H_q(S^q\times S^q)$.

IV.3.10 Lemma. *Let $\bar g$ be generator of $\pi_{q+1}(S^{q+1})$, and let $\{\omega\}=m\partial_0(\bar g)$, $\varphi'=\varphi_\omega$. Then*

$$\varphi'_{0*}(g_1')=\varphi_{0*}(g_1')+2m\varphi_{0*}(g_2')$$
$$\varphi'_{0*}(g_2')=\varphi_{0*}(g_2')\,.$$

Proof. Recall that lemma (IV.1.9) says that the composition

$$\pi_{q+1}(S^{q+1})\xrightarrow{\partial_0}\pi_q(SO(q+1))\xrightarrow{p_*}\pi_q(S^q)$$

is multiplication by 2 if q is odd. Now φ_0' is represented by the composition

$$S^q\times S^q\xrightarrow{\omega'} S^q\times S^q\xrightarrow{\varphi_0} M_0\,,$$

where $\omega': S^q\times S^q\to S^q\times S^q$ is given by $\omega'(x,y)=(x,y\omega(x))$. If y = base point $y_0\in S^q$, then by definition, $y_0\omega(x)=p\omega(x)$, $p:SO(q+1)\to S^q$ is the projection of the bundle. Hence on $S^q\times y_0$, $\varphi_0'(x,y_0)=\varphi_0(x,p\omega(x))$ so $\varphi_0'=\varphi_0(1\times p\omega)\Delta$ on $S^q\times y_0$, $\Delta: S^q\to S^q\times S^q$ given by $\Delta(x)=(x,x)$.

Now $\Delta_*(g)=g_1+g_2$, where now $g\in\pi_q(S^q)$ is the generator, $g_j=(i_j)_*g$, $i_1(x)=(x,y_0)$, $i_2(x)=(y_0,x)$, so that $h(g_i)=g_i'$, h the Hurewicz homomorphism. Then

$$\begin{aligned}\varphi'_{0*}(g_1)&=\varphi_{0*}(1\times p\omega)_*\Delta_*(g)\\&=\varphi_{0*}(1\times p\omega)_*(g_1+g_2)=\varphi_{0*}(g_1+2mg_2)\\&=\varphi_{0*}(g_1)+2m\varphi_{0*}(g_2).\end{aligned}$$

On $y_0\times S^q$, $\omega(y_0)=$ identity of $SO(q+1)$, so $\varphi'_0|y_0\times S^q=\varphi_0|y_0\times S^q$, so $\varphi'_{0*}(g_2)=\varphi_{0*}(g_2)$. The result in homology follows by applying h. □

Returning to diagram (IV.3.2) where $d(\lambda)=\varphi_{0*}(1\otimes[S^q])=h\varphi_{0*}(g_2)$ and $d'(\lambda')=h\varphi_{0*}(g_1)$, if we take the analogous diagram using φ_ω instead of φ, we find $d_\omega(\lambda)=h\varphi_{\omega 0*}(g_2)=d(\lambda)$ and

$$d'_\omega(\lambda')=h\varphi_{\omega 0*}(g_1)=d'(\lambda')+2md(\lambda),$$

or $d(\lambda)=d_\omega(\lambda)$, $d'(\lambda')=d'_\omega(\lambda')-2md_\omega(\lambda)$. Hence (IV.3.4) becomes

IV.3.11

$$s(d'_\omega(\lambda')-2md_\omega(\lambda))+td_\omega(\lambda)=0$$

or

$$sd'_\omega(\lambda')+(t-2ms)\,d_\omega(\lambda)=0.$$

IV.3.12 Proposition. *Let p be prime and let $x\in K_q(M)$ be an element of finite order such that $(x)_p\neq 0$ in $K_q(M;\mathbb{Z}_p)$, where $(\ )_p$ denotes reduction* mod p. *Let $\varphi: S^q\times D^{q+1}\to \operatorname{int}M$ be an embedding which represents x, i.e. $\varphi_*(\mu)=x$, and which defines a normal surgery of (f,b). Then one may choose $\omega: S^q\to SO(q+1)$ so that $\varphi_\omega: S^q\times D^{q+1}\to\operatorname{int}M$ also defines a normal surgery of (f,b), order (torsion $K_q(M'_\omega))\leqq$ order(torsion $K_q(M)$), and* $\operatorname{rank}_{Z_p}K_q(M'_\omega;\mathbb{Z}_p)<\operatorname{rank}_{Z_p}K_q(M;\mathbb{Z}_p)$.

Proof. By (IV.3.2), $H_q(M)/(x)\cong H_q(M')/(y)$, where (x) indicates the subgroup generated by x, etc. By (IV.3.5), order $x=s$, and order $y=t$ if $t\neq 0$, order $y=\infty$ if $t=0$, where (IV.3.4) gives $sd'(\lambda')+td(\lambda)=0$. By (IV.3.9), we may change φ so that (IV.3.4) becomes (IV.3.11) $sd'_\omega(\lambda')+(t-2ms)\,d_\omega(\lambda)=0$, so that $H_q(M)/(x)\cong H_q(M'_\omega)/(y_\omega)$ and order $y_\omega=t-2ms$ if $t-2ms\neq 0$ and order $y_\omega=\infty$ if $t-2ms=0$. Choose m so that $-s\leqq(t-2ms)\leqq s$, so that order $y_\omega\leqq$ order x or order $y_\omega=\infty$. Hence order torsion $H_q(M'_\omega)\leqq$ order torsion $H_q(M)$. Hence

$$\text{order torsion}\,K_q(M'_\omega)\leqq\text{order torsion}\,K_q(M).$$

But if $(x)_p\neq 0$, then by (IV.2.16) $\operatorname{rank}_{Z_p}K_q(M'_\omega;\mathbb{Z}_p)<\operatorname{rank}_{Z_p}K_q(M;\mathbb{Z}_p)$. □

Now we complete the proof of (IV.2.1) for $m=2q+1$, q odd.

Let (f,b) be a normal map, and by (IV.3.1) we may make f q-connected, and $K_q(M)$ a torsion group. Let p be the largest prime dividing order $K_q(M)$, and let $x \in K_q(M)$ be an element such that $(x)_p \neq 0$ in $K_q(M;\mathbb{Z}_p)$. By Whitney's embedding theorem we may find an embedded $S^q \subset \operatorname{int} M^{2q+1}$ representing x, and by (IV.1.6) and (IV.1.12) we may extend this embedding to an embedding $\varphi: S^q \times D^{q+1} \to \operatorname{int} M$ such that φ defines a normal surgery on (f, b). By (IV.3.12) φ may be chosen so that order(torsion $K_q(M')) \leqq$ order(torsion $K_q(M)$) and

$$\operatorname{rank}_{Z_p} K_q(M';\mathbb{Z}_p) < \operatorname{rank}_{Z_p} K_q(M;\mathbb{Z}_p).$$

Proceeding in this fashion step by step we will find after a finite number of such surgeries, a normal cobordism of (f, b) to (f_1, b_1) such that f_1 is q-connected, order (torsion $K_q(M_1)) \leqq$ order (torsion $K_q(M)$), and $\operatorname{rank}_{Z_p} K_q(M_1;\mathbb{Z}_p) = 0$. By (I.2.8), $K_q(M_1;\mathbb{Z}_p) \cong K_q(M_1) \otimes \mathbb{Z}_p$ since $K_i(M_1) = 0$ for $i < q$, and it follows that $K_q(M_1)$ is a torsion group of order prime to p, and order $K_q(M_1) \leqq$ order $K_q(M)$. Since $K_q(M)$ has p-torsion, it follows that order $K_q(M_1) <$ order $K_q(M)$. Hence we have reduced the order of the kernel, and so by a finite sequence of these steps we may make the kernel 0, thus obtaining a normal cobordism of (f, b) with $(\bar{f}, \bar{b})$, where $\bar{f}$ is q-connected, and $K_q(\bar{M}) = 0$. Hence $\bar{f}$ is $(q+1)$-connected, and (IV.2.1) is proved for $m = 2q+1$, q odd. □

This completes the proof of (IV.2.1) for m odd.

§ 4. Proof of the Fundamental Theorem for m even

If $m = 2q$, (f, b) a normal map $f: (M, \partial M) \to (A, B)$,

$$(f \mid \partial M)_*: H_*(\partial M) \to H_*(B)$$

an isomorphism, and f is q-connected, then $K_i(M) = 0$ for $i < q$ and by Poincaré duality $K^{m-i}(M, \partial M) \cong K^{m-i}(M) = 0$ for $i < q$ (see (I.2.6)). By the Universal Coefficient property (I.2.8), it follows that $K_i(M) = 0$ for $i > q$, and $K_q(M)$ is free. Let $x \in K_q(M)$ be represented by an embedding $\alpha: S^q \to \operatorname{int} M$, so that $(\beta, \alpha) \in \pi_{q+1}(f)$,

$$\begin{array}{ccc} S^q & \xrightarrow{\alpha} & M \\ \downarrow & & \downarrow \\ D^{q+1} & \xrightarrow{\beta} & A \end{array}$$

and define $\bar{M} = M \bigcup_\alpha D^{q+1}$, $\bar{f}: \bar{M} \to A$ extending f, defined by β. By (IV.1.6) there is an obstruction $\mathcal{O} \in \pi_q(V_{k,q}) = \mathbb{Z}$ if q is even, $\mathbb{Z}_2$ if q is odd, such that $\mathcal{O} = 0$ if and only if $\bar{f}: \bar{M} \to A$ can be thickened to a normal cobordism. Let $x' \in K^q(M, \partial M)$, such that $[M] \cap x' = x \in K_q(M)$.

Recall that in Chapter III we defined a bilinear pairing (,) on $K^q(M, \partial M)$ and a quadratic form $\psi: K^q(M, \partial M;\mathbb{Z}_2) \to \mathbb{Z}_2$.

IV.4.1 Theorem. *The obstruction $\mathcal{O}$ above for thickening $\bar{f}: \bar{M} \to A$ to a normal cobordism is given by*

$$\mathcal{O} = (x', x') \quad \text{if } q \text{ is even},$$

$$\mathcal{O} = \psi((x')_2) \quad \text{if } q \text{ is odd},$$

where $(\ )_2$ *denotes reduction* mod 2.

Assuming (IV.4.1) for the moment we will complete the proof of (IV.2.1), i.e. in the case m even.

If (f, b) is normally cobordant rel B to a homotopy equivalence then it follows from (II.1.1) that $\sigma(f, b) = 0$.

Let us assume then that $\sigma(f, b) = 0$ and show how to construct a normal cobordism of (f, b) to a homotopy equivalence.

First suppose $m = 2q$ and q is even. Then $\sigma(f, b) = \frac{1}{8} I(f)$, so if $\sigma(f, b) = 0$, then $I(f) =$ signature of $(\ ,\ )$ on $K^q(M, \partial M) = 0$. By (IV.1.13), we may assume that $K^i(M) \cong K^i(M, \partial M) = 0$ for $i < q$, and free for $i = q$. By (III.1.3), there is an $x' \in K^q(M, \partial M)$ such that $(x', x') = 0$, so by (IV.4.1), $[M] \cap x' = x \in K_q(M)$ can be represented by $\varphi: S^q \times D^q \to \operatorname{int} M$, $\varphi_*(\mu) = x$, μ generator of $H_q(S^q \times D^q)$, and the surgery based on φ defines a normal cobordism of (f, b). But we may choose x' to be indivisible, i.e. a generator of a direct summand of $K^q(M, \partial M)$. Hence, by (IV.2.14)

$$\operatorname{rank} K_q(M') < \operatorname{rank} K_q(M),$$

f' still q-connected, where M', $f': (M', \partial M') \to (A, B)$ is the result of the normal surgery based on φ (Actually (IV.3.2) shows that the rank goes down by 2). Since (f, b) and (f', b') are normally cobordant, $I(f') = I(f) = 0$, and we proceed in this fashion until K_q has been reduced to zero and we get a $(q+1)$-connected map.

If $m = 2q$, q odd, the $\sigma(f, b) = c(f, b) =$ Arf invariant of ψ on $K^q(M, \partial M; \mathbb{Z}_2)$. If $\sigma(f, b) = 0$, from (III.1.8) for example, we may deduce the existence of $y \in K^q(M, \partial M; \mathbb{Z}_2)$ with $\psi(y) = 0$. If f is q-connected, then $K^q(M, \partial M; \mathbb{Z}_2) = K^q(M, \partial M) \otimes \mathbb{Z}_2$, and $y = (x')_2$ for some $x' \in K^q(M, \partial M)$, x' indivisible. By (IV.4.1), $x = [M] \cap x'$ is represented by $\varphi: S^q \times D^q \to \operatorname{int} M$, φ defining a normal cobordism, and by (IV.2.14), $\operatorname{rank} K_q(M') < \operatorname{rank} K_q(M)$, f' still q-connected. Also $\sigma(f', b') = \sigma(f, b) = 0$ since (f', b') and (f, b) are normally cobordant, so we may proceed as above till we obtain a $(q+1)$-connected map. This finishes the proof of (IV.2.1) and the Fundamental Theorem. □

It remains then only to prove (IV.4.1), to which we devote the rest of this chapter.

Let (f, b) be a normal map $f: (M, \partial M) \to (A, B)$, $\dim M = m = 2q$, and suppose f is q-connected. Let $x \in K_q(M)$ be represented by an embedding $\alpha: S^q \to \operatorname{int} M$, let ζ^q be the normal bundle of $\alpha(S^q)$ in M, and let

$\bar{M} = M \bigcup_{\alpha} D^{q+1}$, $\bar{f} : \bar{M} \to A$ an extension of f. Let $\mathcal{O} \in \pi_q(V_{k,q})$ be the obstruction to thickening $\bar{M}$ and $\bar{f}$ to a normal cobordism (see (IV.1.6)), and let $\partial : \pi_q(V_{k,q}) \to \pi_{q-1}(SO(q))$ be the boundary in the homotopy exact sequence of the fibre bundle $p : SO(k+q) \to V_{k,q} = SO(k+q)/SO(q)$ with fibre $SO(q)$.

IV.4.2 Proposition. *$\partial \mathcal{O}$ = characteristic map of $\zeta \in \pi_{q-1}(SO(q))$.*

Proof. Let $x_0 \in S^q$ be a base point so that if $h : S^q \to SO(q+k)/SO(q) = V_{k,q}$ represents $\mathcal{O}$, $h(x_0) = p(\mathcal{F}_0)$, where $p : SO(q+k) \to V_{k,q}$ is projection and $\mathcal{F}_0 \in SO(q+k)$ is the base point, a $(k+q)$-frame in R^{k+q}, $p(\mathcal{F})$ = the first k elements of the $(k+q)$ frame $\mathcal{F}$, so that $p(\mathcal{F})$ is a k-frame. Divide S^q into two cells, $S^q = D^q_+ \cup D^q_-$, $x_0 \in D^q_+ \cap D^q_- = S^{q-1} = \partial D^q_+ = \partial D^q_-$. We may assume that $h(D^q_-) = p(\mathcal{F}_0)$, since D^q_- is contractible. Let $\hat{h} : D^q_+ \to SO(q+k)$ such that $\hat{h}(x_0) = \mathcal{F}_0$ and $p\hat{h} = h$ on D^q_+. Then $p\hat{h}(S^{q-1}) = h(S^{q-1}) = p(\mathcal{F}_0)$, so that the first k elements of $\hat{h}(y)$ for $y \in S^{q-1}$ are the base frame of $V_{k,q}$. Hence there is a map $\gamma : S^{q-1} \to SO(q)$ such that $\hat{h}(y) = \mathcal{F}_0(i\gamma(y))$, where $i : SO(q) \to SO(q+k)$ is the representation of $SO(q)$ acting on the subspace of R^{q+k} orthogonal to the space spanned by $p(\mathcal{F}_0)$. Then γ represents $\partial \mathcal{O} \in \pi_{q-1}(SO(q))$, by the definition of ∂ (see [60]).

Now ζ is the orthogonal bundle to the trivial bundle spanned by $h(x)$, for $x \in S^q$. Since $h(D^q_-) = p(\mathcal{F}_0)$, the last q vectors in $\mathcal{F}_0$ give a trivialization of ζ over D^q_-, and since $p\hat{h} = h$, the last q vectors of $\hat{h}(x)$, $x \in D^q_+$, give a trivialization of ζ over D^q_+. Since $\gamma(y)$ for $y \in S^{q-1}$ sends the last part of $\mathcal{F}_0$ into the last part of $\hat{h}(y)$, it follows that γ is the characteristic map of ζ (see [60; (18.1)]). □

Now from results of IV., § 1 we may derive easily

IV.4.3 Proposition. *The boundary $\partial : \pi_q(V_{k,q}) \to \pi_{q-1}(SO(q))$ is a monomorphism for $q \neq 1$, 3 or 7.*

Proof. Considering the inclusion of total spaces, $SO(q+1)$ in $SO(q+k)$, and the projection $SO(q+k) \to SO(q+k)/SO(q-1)$, we get the commutative diagram where p_1, p_2 and p_3 are projections of fibre bundles, i_1, i_2 and i_3 inclusions of the fibres:

$$\begin{array}{ccccc}
SO(q) & \longrightarrow & SO(q) & \xrightarrow{p'} & SO(q)/SO(q-1) = S^{q-1} \\
\downarrow{\scriptstyle i_1} & & \downarrow{\scriptstyle i_2} & & \downarrow{\scriptstyle i_3} \\
SO(q+1) & \xrightarrow{j} & SO(q+k) & \longrightarrow & V_{k+1,q-1} \\
\downarrow{\scriptstyle p_1} & & \downarrow{\scriptstyle p_2} & & \downarrow{\scriptstyle p_3} \\
S^q = V_{1,q} & \xrightarrow{j'} & V_{k,q} & \longrightarrow & V_{k,q}\,.
\end{array}$$

Let ∂_i, $i=1,2,3$ be the boundary operators associated with the bundle projections p_i. By (IV.1.9), if q is even, then $p'_*\partial_1:\pi_q(S^q)\to\pi_{q-1}(S^{q-1})$ is multiplication by 2, hence is a monomorphism. But by commutativity of the diagram, $p'_*\partial_1=\partial_3 j'_*$. Hence j'_* is a monomorphism, and since by (IV.1.12) $\pi_q(V_{k,q})=\mathbb{Z}$ if q is even, it follows that $\partial_3=\partial$ is a monomorphism if q is even.

If $q\neq 1$, 3 or 7, q odd, then by (IV.1.11) $\ker i_*=\mathbb{Z}_2$, where $i_*:\pi_{q-1}(SO(q))\to\pi_{q-1}(SO(q+1))$. Hence ∂_1 is onto $\mathbb{Z}_2\subset\pi_{q-1}(SO(q))$ and since $j'_*:\pi_q(S^q)\to\pi_q(V_{k,q})$ is onto by (IV.1.12), $\partial_1=\partial_3 j'_*$, it follows that $\partial_3(\pi_q(V_{k,q}))\supset\mathbb{Z}_2$. Since $\pi_q(V_{k,q})=\mathbb{Z}_2$ for q odd by (IV.1.12), it follows that $\partial_3=\partial$ is a monomorphism for $q\neq 1$, 3 or 7. □

Thus for $q\neq 1$, 3 or 7, the obstruction $\mathcal{O}$ to doing normal surgery on $S^q\subset M^{2q}$ can be identified with the characteristic map of ζ, its normal bundle in M, $\mathcal{O}\in\ker i_*\subset\pi_{q-1}(SO(q))$, and is therefore 0 if ζ is trivial. Now $\ker i_*$ is generated by $\partial_1(i)$, where $i\in\pi_q(S^q)$ is the class of the identity, so that $\partial_1(i)$ is the characteristic map for the tangent bundle τ of S^q. It follows that $\mathcal{O}=\lambda(\partial_1(i))$, some $\lambda\in\mathbb{Z}$.

Now if q is even the Euler class $\chi(\tau)=2g\in H^q(S^q)$, where g is the generator such that $g([S^q])=1$. This follows from the general formula $\chi(\tau_M)=\chi(M)g$, or may be deduced for $M=S^q$, q even, using the fact that τ_M is equivalent to the normal bundle of the diagonal M in $M\times M$. For if $U\in H^q(E,E_0)$ is the Thom class, it follows from (IV.2.8) that

$$[S^q\times S^q]\cap\eta^* U=[S^q]\otimes 1+1\otimes[S^q]$$

the homology class of the diagonal, where $\eta:S^q\times S^q\to E/E_0$ is the natural collapse. Hence $\eta^* U=g\otimes 1+1\otimes g$, and

$$\eta^*(U^2)=(\eta^* U)^2=(g\otimes 1+1\otimes g)^2=2g\otimes g\,,$$

if q is even. Since η^* is an isomorphism on H^{2q}, it follows that $U^2=2gU$, so $\chi(\tau)=2g$, since by definition $\chi(\xi)U_\xi=(U_\xi)^2$, for a bundle ξ.

Now the Euler class is represented by the universal Euler class $\chi\in H^q(BSO(q))$, where $BSO(q)$ is the classifying space for oriented q plane bundles (see [60] or [32]). That is, if $c:X\to BSO(q)$ is the classifying map of a q-plane bundle ξ over X, $c^*(\gamma)=\xi$, where γ is the universal q-plane bundle over $BSO(q)$, then $\chi(\xi)=c^*(\chi)$. If $c:S^q\to BSO(q)$ represents τ_{S^q}, then $c^*(\chi)=2g$, as above, but if $c':S^q\to BSO(q)$ represents $\lambda(\tau_{S^q})$ in the homotopy group $\pi_{q-1}(SO(q))$, then λc and c' are homotopic, i.e. $\{\lambda c\}=\{c'\}$ in $\pi_q(BSO(q))$. Hence $c'^*=\lambda c^*$, so we get:

IV.4.4 Lemma. *If q is even and $\partial_2\mathcal{O}=\lambda\partial_1(i)$, then $\chi(\zeta)=2\lambda g$, where ζ is the normal bundle of αS^q in M^{2q}, representing an element in $K_q(M)$, $\mathcal{O}$ the obstruction to doing a normal surgery on S^q.* □

IV.4.5 Lemma. *$\chi(\zeta)[S^q]=(x',x')$, where $[M]\cap x'=x$, $\alpha: S^q\to M^{2q}$ is an embedding representing $x\in K_q(M)$, ζ the normal bundle of $\alpha(S^q)$, as above.*

Proof. $\chi(\zeta)U=U^2$ by the definition of χ, where $U\in H^q(E(\zeta)/E_0(\zeta))$ is the Thom class. Clearly $(\chi(\zeta))[S^q]=(\chi(\zeta)U)[E]=U^2[E]=(\eta^* U)^2[M]$, where $[E]\in H_{2q}(E(\zeta)/E_0(\zeta))$ is the orientation class, so $[E]=\eta_*[M]$, where $\eta: M/\partial M\to E/E_0$ is the natural collapse.

By (IV.2.8), $[M]\cap\eta^* U=x$, so that $\eta^* U=x'$. Hence

$$\chi(\zeta)[S^q]=(\eta^* U)^2[M]=(x')^2[M]=(x',x'). \quad \square$$

By (IV.4.4) and (IV.4.5), for q even, $(x',x')=2\lambda$ where $\partial_2\mathcal{O}=\lambda\partial_1(i)$. By (IV.4.3) ∂_2 is a monomorphism for q even, so we may identify $\mathcal{O}$ with (x',x'), which proves the first part of (IV.4.1), i.e. for q even.

The result for q odd is more delicate since it is not dependent only on the normal bundle ζ of S^q in M^{2q} if $q=1,3$ or 7, and even for $q\neq 1,3$ or 7, it is more difficult to detect the normal bundle ζ.

Let $\alpha_i: S^q\to M^{2q}$, $i=1,2$ be embeddings representing $x_i\in K_q(M)$, where as usual (f,b) is a normal map $f:(M,\partial M)\to(A,B)$,

$$(f\mid\partial M)_*: H_*(\partial M)\to H_*(B)$$

an isomorphism. Suppose $\alpha_1(S^q)\cap\alpha_2(S^q)=\emptyset$, and let $\mathcal{O}_1$ and $\mathcal{O}_2$ be the obstructions to doing normal surgery on $\alpha_1(S^q)$ and $\alpha_2(S^q)$ respectively. Join $\alpha_1(S^q)$ and $\alpha_2(S^q)$ be an arc in the complements, and by thickening this arc to a tube $T=D^q\times[1,2]$ we take

$$(\alpha_1(S^q)-(D^q\times 1))\cup\partial_0 T\cup(\alpha_2(S^q)-(D^q\times 2))$$

where $\partial_0 T=\partial D^q\times[1,2]$, $D^q\times i=T\cap\alpha_i S^q$. This gives us an embedding $\alpha: S^q\to M^{2q}$ representing x_1+x_2, which can be made differentiable by "rounding the corners."

IV.4.6 Lemma. *$\mathcal{O}=\mathcal{O}_1+\mathcal{O}_2$ in $\pi_q(V_{k,q})$.*

Proof. If we thicken the embedding of $T\subset M$ by multiplying by $[0,\varepsilon]$ we get an embedding of $T\times[0,\varepsilon]$ in $M\times I$ and if we have $M\subset D^{m+k}$, $M\times I\subset D^{m+k}\times I$, this gives us $T\times[0,\varepsilon]\subset D^{m+k}\times I$, and if $\alpha_i S^q=\partial D_i^{q+1}$, $D_i^{q+1}\subset D^{m+k}\times I$ meeting $D^{m+k}\times 0$ transversally in $\alpha_i(S^q)$, so then we may assume a neighborhood of $\alpha_i(S^q)$ in D_i^{q+1} is just $\alpha_i(S^q)\times[0,\varepsilon]$. If we take

$$\begin{aligned} D^{q+1}=D_1^{q+1}&-(D^q\times 1\times[0,\varepsilon])\cup(\partial D^q\times[1,2]\times[0,\varepsilon])\\ &\cup(D^q\times[1,2]\times\varepsilon)\cup D_2^{q+1}-(D^q\times 2\times[0,\varepsilon]),\end{aligned}$$

then this is a $(q+1)$ cell meeting $D^{m+k}\times 0$ transversally in $\alpha(S^q)$, and we may smooth this D^{q+1} and $\alpha(S^q)$ together by "rounding corners."

The resulting D^{q+1} is the union of three cells, $D^{q+1}=A_1\cup B\cup A_2$, where $A_i=D_i^{q+1}-\operatorname{int}C_i$ where C_i is a $(q+1)$-cell, $\partial C_i\cap\partial D_i=F_i$, F_i a q-cell in ∂D_i, $B\cap A_i=\partial C_i\cap A_i\subset\partial B$, and

$$\partial B-(\partial C_1\cap A_1)-(\partial C_2\cap A_2)=S^{q-1}\times I\,.$$

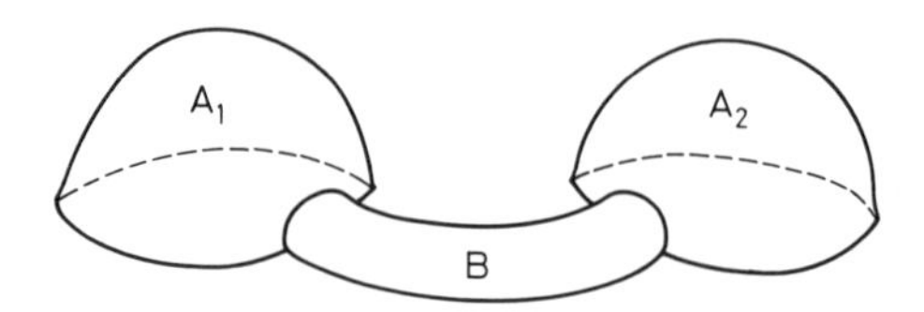

Since the definition of the obstruction $\mathcal{O}$ does not depend on the choice of the framing of the normal bundle γ of the disk D^{q+1}, we may assume that the framings over D^{q+1}, D_1^{q+1} and D_2^{q+1} have been chosen so that the framings over D^{q+1} and D_i^{q+1} coincide over A_i, $i=1,2$. Further we may assume that the framings of v, the normal bundle of M in D^{m+k} over αS^q, $\alpha_1 S^q$, $\alpha_2 S^q$, induced by b, have been chosen so that over F_i they are all the same, coming from a framing of $v|T$, (T is a cell), and the framings of γ, γ_1 and γ_2 may be assumed to extend that of v over $T\cap\alpha S^q$, $T\cap\alpha_i S^q$, $i=1,2$. Thus the three maps β, β_i, $i=1,2$, $\beta:\alpha S^q\to V_{k,q}$, etc. defining $\mathcal{O}$, $\mathcal{O}_i$, $i=1,2$, may be taken to be the base k-frame over $T\cap\alpha S^q$, $T\cap\alpha_i S^q$, $i=1,2$, and $\beta|\alpha_i(S^q)\cap\alpha(S^q)=\beta_i|\alpha_i(S^q)\cap\alpha(S^q)$. It follows that for the homotopy classes $\{\beta\}=\{\beta_1\}+\{\beta_2\}$ in $\pi_q(V_{k,q})$ or $\mathcal{O}=\mathcal{O}_1+\mathcal{O}_2$. □

IV.4.7 Lemma. *If $\mathcal{O}=0$, then $\psi((x')_2)=0$, with notation as above.*

Proof. Since $\mathcal{O}=0$, we have a normal surgery based on $\alpha:S^q\to M^{2q}$, so that the trace is a normal cobordism W^{2q+1}, $\partial W=M\cup(\partial M\times I)\cup M'$, and if $i:\partial W\to W$ is inclusion, $i_*k_*x=0$, where $k:M\to\partial W$. It follows from (I.2.7), that $x''=i^*z$, $z\in K^q(W)$, where $[\partial W]\cap x''=k_*x$, $x''\in K^q(\partial W)$ and $K^q(W)$ is defined by the map of $F:W\to A\times I$ extending f on M.

It follows from (III.4.13) that $\psi_0((i^*z)_2)=\psi_0((x'')_2)=0$, where ψ_0 is defined on $K^q(\partial W;\mathbb{Z}_2)$, for the map $\partial F:\partial W\to A\times 0\cup B\times I\cup A\times 1$. Now ∂F is clearly the sum of (f,b) on M and (f',b') (the result of the surgery) on M'. By (III.4.15) $\psi_0(\eta^*(x')_2)=\psi((x')_2)$, $x'\in K^q(M,\partial M)$, so it remains to show that $\eta^*(x')_2=(x'')_2$, $(\eta:\partial W\to M/\partial M)$.

Consider $\quad k_*x=k_*([M]\cap x')=k_*(\eta_*[\partial W]\cap x')=[\partial W]\cap\eta^*x'$, using identities of the cap product (compare IV.2.8) so that since $[\partial W]\cap x''=k_*x$, it follows that $x''=\eta^*x'$, and hence $\psi((x')_2)=0$. □

Now we prove that $\mathcal{O}=\psi((x')_2)$. If $\mathcal{O}=0$, then $\psi((x')_2)=0$ by (IV.4.7). So it remains to show that if $\mathcal{O}=1$ then $\psi((x')_2)=1$.

By taking the connected sum with the map $S^q\times S^q\to S^{2q}$, or alternatively doing a normal surgery on a $S^{q-1}\subset D^{2q}\subset M^{2q}$, we may add to

$K_q(M)$ the free module on two generators a_1, a_2, corresponding to $[S^q]\otimes 1$ and $1\otimes[S^q]$ in $H^q(S^q\times S^q)$ and add to $K^q(M,\partial M)$ the elements g_1, g_2 such that $[M\#(S^q\times S^q)]\cap g_i=a_i$, with $(g_1,g_2)=1$, $(g_i,g_i)=0$, $i=1,2$, orthogonal to the original $K^q(M,\partial M)$, and $\psi(g_1)=\psi(g_2)=0$. Hence $\psi(g_1+g_2)=\psi(g_1)+\psi(g_2)+(g_1,g_2)=0+0+1=1$. If $\beta: S^q\to M\#(S^q\times S^q)$ represents the diagonal class

$$a_1+a_2\in K_q(M\#(S^q\times S^q)),$$

it follows from (IV.4.7) that the obstruction $\mathcal{O}$ to surgery on β, $\mathcal{O}=1$, since if it were 0 then $\psi(g_1+g_2)$ would be 0. Then on the sum embedding $\alpha+\beta$ representing $x+(a_1+a_2)$ the obstruction $\mathcal{O}''=\mathcal{O}+\mathcal{O}'$ by (IV.4.6), so that $\mathcal{O}''=1+1=0$. Hence $\psi((x')_2+(g_1+g_2))=0$ by (IV.4.7). But since $((x')_2,(g_1+g_2))=0$,

$$\psi((x')_2+(g_1+g_2))=\psi((x')_2)+\psi(g_1+g_2)=\psi((x')_2)+1=0,$$

so that $\psi((x')_2)=1$. $\square$

V. Plumbing

In this chapter we will describe the process of "plumbing" introduced by Milnor [47], which constructs manifolds with prescribed quadratic forms in the middle dimension. See Hirzebruch [31] for another discussion.

§ 1. Intersection

In this paragraph we review the intersection theory of submanifolds of a manifold.

Let N_1^p, N_2^q be smooth submanifolds of a smooth manifold M^m, $m=p+q$. (Smoothness may of course be replaced by much weaker conditions in what follows.) A point $x\in N_1\cap N_2$ will be called a discrete point if x has an open neighborhood V in M such that $V\cap N_1\cap N_2=x$. If every point of $N_1\cap N_2$ is discrete, then $N_1\cap N_2$ is a discrete subset of M.

If $x\in N_1\cap N_2$ is discrete and V is open in M such that $V\cap N_1\cap N_2=x$, then $(V-N_1)\cup(V-N_2)=(V-x)$. Hence we have a pairing given by relative cup product

V.1.1

$$H^q(V, V-N_1)\otimes H^p(V, V-N_2)\to H^{p+q}(V, V-x)\,.$$

Now suppose M, N_1 and N_2 are oriented, so that for each point $x\in M$ we are given a generator $[M]_x\in H_m(M, M-x)$, (for $y\in N_1$ we are given $[N_1]_y\in H_p(N_1, N_1-y)$, for $z\in N_2$ we are given $[N_2]_z\in H_q(N_2, N_2-z)$), in a compatible way for all the points $x\in M$ ($y\in N_1$, $z\in N_2$), (see the proof of (I.3.5)). Let E_i, $i=1,2$ be a tubular neighborhood of N_i in M, $E_i^0=$ the complement of the zero cross-section. Then the inclusion

$$(E_i, E_i^0)\subset(M, M-N_i)$$

is an excision, so $H^*(M, M-N_i)\cong H^*(E_i, E_i^0)$. By the Thom isomorphism theorem if E_i are oriented there are elements $U_1\in H^q(E_1, E_1^0)$, $U_2\in H^p(E_2, E_2^0)$ such that r^*U_i is a generator of $H^*(V, V-N_i)$, for any small neighborhood such that $V=A\times B$, B a ball in N_i, $V\cap N_i=0\times B$,

r denotes inclusion and such that $\cup U_i$ and $\cap U_i$ induce isomorphisms, (see (I.4.3), (II.2.3), (II.2.6)). Let us assume all orientations are chosen compatible so that $[M]_x \cap r^* U_i = [N_i]_x$ for $x \in N_i$.

Then we may define the sign or orientation of a discrete point $x \in N_1 \cap N_2$ by

V.1.2

$$\operatorname{sgn}(x) = (r^* U_1 \cup r^* U_2)\,[M]_x,$$

(using (V.1.1)).

We shall call x a (homologically) transversal point of intersection if $\operatorname{sgn}(x) = \pm 1$. This will obviously be the case for transversal intersections in the usual geometrical sense.

Recall the definition of the intersection of homology classes (see IV § 2). Let M^m be a compact oriented manifold with boundary, and let $x \in H_p(M)$, $y \in H_q(M, \partial M)$, $p+q=m$. Define $x \cdot y = (x', y') = (x' \cup y')\,[M]$, where $x' \in H^{m-p}(M, \partial M)$, $y' \in H^{m-q}(M)$, such that $[M] \cap x' = x$, $[M] \cap y' = y$. The same definition also works for $x, y \in H_*(M)$, i.e. $x \cdot y = x \cdot j_* y$.

Suppose N_1^p, N_2^q are compact oriented submanifolds of M^m, a compact oriented manifold with boundary, $m = p+q$, and suppose N_1 is closed and $\partial N_2 \subset \partial M$, and $\partial M \cap N_1 = \emptyset$, $\partial M \cap N_2 = \partial N_2$. Let N_1, N_2 and M be oriented, and suppose N_1 intersects N_2 (homologically) transversally, and let $i_j : N_j \to M$ be the inclusions $j = 1, 2$.

V.1.3 Theorem.

$$(i_{1*}[N_1]) \cdot (i_{2*}[N_2]) = \sum_{x \in N_1 \cap N_2} \operatorname{sgn}(x).$$

In other words the intersection of the homology classes counts the number of intersection points, with the sign.

Proof. Let $U_1 \in H^{m-p}(E_1, E_1^0)$, $U_2 \in H^{m-q}(E_2, E_2^0)$ be the Thom classes of the normal bundles of N_1 and N_2 (notation as above). Let $k_i : (E_i, E_i^0) \to (M, M - N_i)$ be inclusions (excisions), $j : (M, \partial M) \to (M, M - N_1)$, $l : M \to (M, M - N_2)$. Let $u_i \in H^*(M, M - N_i)$ be such that $k_i^* u_i = U_i$, and let $x_1 = j^* u_1$, $x_2 = l^* u_2$. Then

V.1.4

$$[M] \cap x_j = i_{j*}[N_j], \quad j = 1, 2.$$

For by (IV.2.7), $i_{1*}([N_1]) = i_{1*}([E] \cap U_1) = [M] \cap j^* u_1 = [M] \cap x_1$, (similar proof for $j = 2$). Hence, by definition $(i_{1*}[N_1]) \cdot (i_{2*}[N_2]) = (x_1 \cup x_2)\,[M]$. But $x_1 \cup x_2 = (j^* u_1) \cup (l^* u_2) = h^*(u_1 \cup u_2)$, where

$$h : (M, \partial M) \to (M, (M - N_1) \cup (M - N_2)).$$

Hence $(x_1 \cup x_2)[M] = (h^*(u_1 \cup u_2))[M] = (u_1 \cup u_2)(h_*[M])$. Now

$$h_*[M] = q_* \sum_{x \in N_1 \cap N_2} [M]_x \,,$$

$$q: \bigcup (V_x, V_x - x) \to (M, M - (N_1 \cap N_2)),$$

$$(x_1 \cup x_2)[M] = \sum_{x \in N_1 \cap N_2} (u_1 \cup u_2)(q_*[M]_x) \,.$$

Since $k_i^* u_i = U_i$, it follows that

$$(u_1 \cup u_2)(q_*[M]_x) = ((r^* U_1) \cup (r^* U_2))[M]_x = \operatorname{sgn}(x) \,,$$

and the result follows. □

Now let N^q be a closed submanifold in interior of M^{2q}, and let ζ^q be its normal bundle. If we make N transversal to itself (using for example Thom's transversality theorem), then (V.1.3) implies that $i_*[N] \cdot i_*[N] = \Sigma \operatorname{sgn}(x)$, the summation running over points x common to the two copies of N. However this self-intersection number is also interpretable in terms of ζ:

V.1.5 Proposition. $i_*[N] \cdot i_*[N] = \chi(\zeta)[N]$ (see (IV.4.5)).

Proof. By (IV.2.8), $[M] \cap \bar{g}^* U = i_*[N]$ where $\bar{g}: M/\partial M \to E/\partial E = T(\zeta)$, $E =$ total space of ζ is a tubular neighborhood of N. Hence

$$i_*[N] \cdot i_*[N] = (\bar{g}^* U)^2 [M] = (U^2)(\bar{g}_*[M]) = (\chi(\zeta) U)([E])$$
$$= \chi(\zeta)([E] \cap U) = \chi(\zeta)[N] \,. \quad □$$

§ 2. Plumbing Disk Bundles

Now we describe the process of "plumbing" disk bundles over manifolds.

Let ζ_i^q be a q-plane bundle of a q-dimensional smooth manifold N_i^q, and let $E_i =$ the total space of the closed disk bundle associated to ζ_i. We shall suppose ζ_i, N_i and E_i oriented compatibly, $i = 1, 2$.

Let $x_i \in N_i^q$, $i = 1, 2$, and let D_i^q be a ball neighborhood of x_i in N_i. Since D_i^q is contractible, ζ_i is a product over D_i^q, so that a neighborhood of x_i in E_i is diffeomorphic to $D_i^q \times D_i^{q'}$, $x \times D_i^{q'}$ being the fibres of E_i. Let $h_+ : D_1^q \to D_2^{q'}$, $k_+ : D_1^{q'} \to D_2^q$, $(h_- : D_1^q \to D_2^{q'})$, $(k_- : D_1^{q'} \to D_2^q)$ be orientation preserving (orientation reversing) diffeomorphisms.

We define the "plumbing" of E_1 with E_2 at x_1 and x_2 by taking $E_1 \cup E_2$ and identifying $D_1^q \times D_1^{q'}$ with $D_2^q \times D_2^{q'}$ by the map $I_+(x, y) = (k_+ y, h_+ x)$ or by the map $I_-(x, y) = (k_- y, h_- x)$, $I_\pm : D_1 \times D_1' \to D_2 \times D_2'$. We shall say we plumb with sign $+1$ if we use I_+, with sign -1 if we use I_-. (Note that we could have used only one manifold, plumbing together 2 points on it, i.e. taking $E_1 = E_2$.) We denote the result of the plumbing by $E_1 \square E_2$, which can be made differentiable by straightening the angles. Since I_+

and I_- are both orientation preserving if q is even (reversing if q is odd), $E_1 \square E_2$ can be oriented compatibly with N_1, ζ_1, N_2 and ζ_2 if q is even (with N_1, ζ_1, $-N_2$, ζ_2 if q is odd). Note that $N_i \subset E_i \subset E_1 \square E_2$, and that $N_1 \cap N_2 = x_1 = x_2$, which is a transversal intersection, and that $\operatorname{sgn}(x_1) = +1$ if we used I_+, $\operatorname{sgn}(x_1) = -1$ if we used I_-.

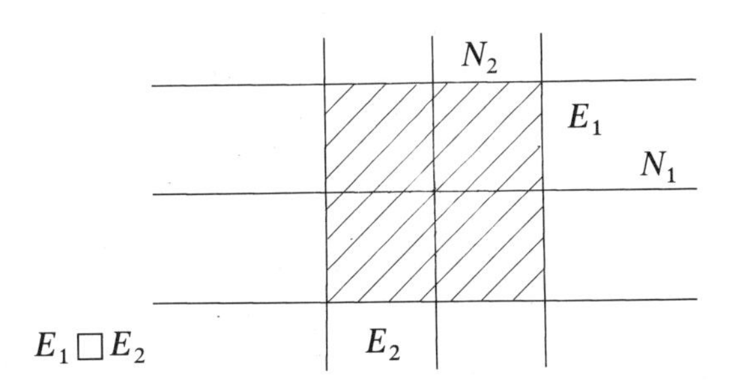

If we choose several different points in N_1 and N_2, then we may plumb E_1 and E_2 together at several different points simultaneously with a prescribed sign at each point. If we plumb at n_{12} points with sign $+1$ for example, we get a manifold $E_1 \square E_2$ with $N_i \subset E_i \subset E_1 \square E_2$ and with $i_{1*}[N_1] \cdot i_{2*}[N_2] = n_{12}$ (see (V.1.3)). If we take a third manifold N_3^q and a q-plane bundle ζ_3^q over N_3, we may plumb E_3 with E_2 at $|n_{23}|$ points, and with E_1 at $|n_{13}|$ points with sign $= \operatorname{sign} n_{ij}$, by simply avoiding the finite number of points of N_2 and N_1 involved with the plumbing of E_1 and E_2, and get a manifold $E_1 \square E_2 \square E_3$ with $N_i \subset E_i \subset E_1 \square E_2 \square E_3$, and $i_*[N_i] \cdot i_*[N_j] = n_{ij}$, where $i \neq j$. (Here we take $n_{ji} = n_{ij}$ if q is even, $n_{ji} = -n_{ij}$ if q is odd.) We may continue this process to plumb together m-different manifolds $E_1, \ldots, E_m$, $E_i =$ total space of ζ_i^q over N_i^q, $i = 1, \ldots, m$, and with prescribed intersections between N_i and N_j for $i \neq j$. The self-intersections of the N_i's are determined by the Euler class $\chi(\zeta_i)$, by (V.1.5).

V.2.1 Theorem. *Let M be an $n \times n$ matrix with integer entries, symmetric, and with even entries on the diagonal. Then for $k > 1$, there is a manifold W^{4k} with boundary such that:*

(i) *W is $(2k-1)$-connected, ∂W is $(2k-2)$ connected, $H_{2k}(W)$ is free abelian and,*

(ii) *the matrix of intersections $H_{2k}(W) \otimes H_{2k}(W) \to \mathbb{Z}$ is given by M (or equivalently the matrix of the bilinear form $(\ ,\ )$ on $H^k(W, \partial W)$),*

(iii) *there is a normal map (f, b), $f: (W, \partial W) \to (D^{4k}, S^{4k-1})$, so that in particular M is the intersection matrix also on $K_{2k}(W)$.*

Proof. Let $M = (m_{ij})$, $i, j = 1, \ldots, n$, $m_{ij} = m_{ji}$, $m_{ii} = 2\lambda_i$, all i, j. Let S_i^q, $i = 1, \ldots, n$ be q-spheres, $q = 2k$, and take ζ_i^q over S_i^q to be $\lambda_i \tau_{S^q}$, i.e. λ_i times τ_{S^q} in the homotopy group $\pi_q(BSO_q)$ or $\pi_{q-1}(SO(q))$. By (IV.4.4),

$\chi(\lambda_i \tau_{S^q}) = 2\lambda_i [S_i^q]$, so by (V.1.5) the intersection number of S_i^q with itself in E_i is $2\lambda_i = m_{ii}$.

Now plumb together the E_i's, $i = 1, \ldots, n$, plumbing E_i with E_j at $|m_{ij}|$ points with sign = sign of m_{ij}. Call the resulting manifold U. We shall say U is the result of plumbing by the matrix M. Then $S_i^q \subset E_i \subset U$ and $i_*[S_i] \cdot i_*[S_j] = m_{ij}$, by the construction.

Since each E_i has S_i^q as a deformation retract, and $E_i \cap E_j = \cup$ (disks), it follows:

V.2.2 Lemma. *If U is the result of plumbing by the matrix M, then U has a deformation retract $\cup S_i^q$, where $S_i^q \cap S_j^q = |m_{ij}|$ points.*

It follows easily that

V.2.3 Each component of U is the homotopy type of a wedge of q-spheres and 1-spheres.

For taking the union of two q-spheres with $(n+1)$ points in common is the homotopy type of $S^q \vee S^q \vee \bigvee_n S^1$. Then (V.2.3) follows by induction.

Now we have constructed U so that it is oriented, and hence τ_U is trivial on the 1-skeleton of U. On each $E_i \subset U$, $\tau_U | E_i = \tau_{S^q} + \lambda_i \tau_{S^q}$, and since $\tau_{S^q} + \varepsilon^1$ is trivial it follows that $\tau_U + \varepsilon^1 | E_i$ is trivial, and hence $\tau_U | S_i^q$ is trivial. It follows from (V.2.2) and (V.2.3) that τ_U is trivial, since it is trivial on each piece of the wedge. It follows that the normal bundle v of $(U, \partial U) \subset (D^{2q+k}, S^{2q+k-1})$ is trivial, k large. Let $b': v \to R^k$ be a framing, R^k = the trivial bundle over a point.

Let $[U] \in H_{2q}(U, \partial U)$ be the orientation class, $\partial[U] \in H_{2q-1}(\partial U)$ the orientation class of ∂U. Take a map $f': \partial U \to S^{2q-1}$ such that $f'^*(g)(\partial[U]) = 1$, $g \in H^{2q-1}(S^{2q-1})$ a generator, and extend f' to $f: (U, \partial U) \to (D^{2q}, S^{2q-1})$, which is possible since D^{2q} is contractible. Let ξ^k be the trivial bundle over D^{2q}, $E(\xi) = D^{2q} \times R^k$ and define $b: v \to \xi$ by $b(v) = (f \pi v, b'(v))$, where $\pi: E(v) \to U$ is projection. This defines a normal map (f, b) as in (V.2.1) (iii).

Now let us look at the effect of plumbing on the boundary of the manifold. We see that for plumbing at one point,

$$\partial(E_1 \square E_2) = (\partial E_1 - D_1^q \times S^{q-1}) \cup (\partial E_2 - D_2^q \times S^{q-1}).$$

Now $\partial E_i - D_i^q \times S^{q-1}$ has the same homotopy type as $\partial E_i - S^{q-1}$, and since codimension $S^{q-1} = q$, $\pi_j(\partial E_i - D_i^q \times S^{q-1}) \to \pi_j(\partial E_i)$ is an isomorphism for $j \leqq q-2$. Also

$$(\partial E_1 - D_1^q \times S^{q-1}) \cap (\partial E_2 - D_2^q \times S^{q-1}) = S^{q-1} \times S^{q-1}$$

which is $(q-2)$ connected.

If E_i is the total space of ζ_i^q over S_i^q, then ∂E_i is $(q-2)$-connected, and it follows that $\partial(E_1 \square E_2 \square \cdots \square E_n)$ is the union of $(q-2)$-connected parts along a large number of $(q-2)$-connected subspaces.

V.2.4 Lemma. *If $q>2$, then for each component X of $E_1 \square E_2 \square \cdots \square E_n$ we have*

(a) $\pi_1(\partial X) \cong \pi_1(X)$ *is free*

(b) $H_i(\partial X) \cong H_i(X) = 0$ *for* $1 < i < q-1$, *where the isomorphisms are induced by inclusion.*

Proof. X is the union of simply connected parts along simply connected subspaces, and similarly for ∂X if $q>2$ by our above remarks. Hence $\pi_1(\partial X)$ and $\pi_1(X)$ are free by van Kampen's theorem. The components of the intersections for the union which gives ∂X are in $1-1$ correspondence with the intersections in the union which gives X, i.e. $S^{q-1} \times S^{q-1} \subset D^q \times D^q$, and hence $\pi_1(\partial X) \cong \pi_1(X)$. This proves (a). Part (b) follows by a similar argument using the Mayer-Vietoris sequence, and the fact that every component involved is $(q-2)$ connected. $\square$

V.2.5 Complement. If $q=2$, $\pi_1(\partial X)$ may not be free, but $\pi_1(X)$ is free and $\pi_1(\partial X) \to \pi_1(X)$ is onto.

The proof is similar.

Now choose an $S^1 \subset \partial X$ which represents a free generator g of the free group $\pi_1(\partial X) \cong \pi_1(X)$. In this low dimension there is no obstruction to doing a normal surgery on S^1, (see (IV.1.6) and (IV.1.12)), so the trace V of the surgery has the homotopy type of $\partial X \cup D^2$, and there exist $\bar{f}: V \to S^{2q-1}$, $\bar{b}: \omega \to \xi$ extending $(f \mid \partial X,\ b \mid \partial X)$, ($\omega =$ normal bundle of V in $D^{m+k} \times I$). Then $X_1 = X \cup V$ along ∂X has the homotopy type of $X \cup D^2$ and hence $\pi_1(X_1) \cong \pi_1(X)/(g)$, $(g) =$ smallest normal subgroup containing g. Since g is a free generator of $\pi_1(X)$, $\pi_1(X_1)$ is free on one less generator, and since $\dim \partial X = 2q-1 > 3$, it follows from (IV.1.2), (IV.1.3) that the same is true for ∂X_1 (where $\partial V = \partial X \cup \partial X_1$) and $\pi_1(\partial X_1) \cong \pi_1(X_1)$. Also it follows easily from the homology sequence of the pair (X_1, X) that $H_i(X) \cong H_i(X_1)$ for $i \neq 1$, and similarly $H_i(\partial X_1) \cong H_i(\partial X)$ for $1 < i < 2q-2$, by a slightly different argument.

The maps (f, b) and $(\bar{f}, \bar{b})$ on X and V fit together to define a new normal map (f_1, b_1), $f_1: (X_1, \partial X_1) \to (D^{2q}, S^{2q-1})$.

Continuing in this way, doing surgeries on circles in the boundary and adding the trace to the manifold, we eventually arrive at an X_n, $\pi_1(X_n) \cong \pi_1(\partial X_n) = 0$, $X \subset X_n$ and $H_i(X_n) \cong H_i(X)$ for $i \neq 1$, $H_i(\partial X_n) \cong H_i(\partial X)$ for $1 < i < 2q-2$. Take the connected sum along the boundaries of these X_n's for all the components and call the result W, so that $U \subset W$.

Then W is connected, $\pi_1(W) = 0$ and $H_i(W) \cong H_i(U)$, $i > 1$ and since $H_i(U) = 0$ for $1 < i < q$ it follows that W is $(q-1)$ connected, and similarly, $H_i(\partial W) \cong H_i(\partial U)$ for $1 < i < q-1$ so that ∂W is $(q-2)$ connected, so (V.2.1)(i) is satisfied. We have constructed normal maps

(f_n, b_n) for each component X_n, so that a normal map is defined on the connected sum along the boundaries, which proves (iii).

Now in $U \subset W$ we have the embedded spheres $S_i^q \subset U^{2q}$ with normal bundle ζ_i^q, and by our construction M is the intersection matrix of the S_i^q. But $i_*[S_i^q]$ give a homology basis of $H_q(U)$, and hence for $H_q(W)$, and the intersection numbers are the same, depending only on a neighborhood of the embedded manifolds S_i^q. Hence M is also the intersection matrix for W, which proves (V.2.1) (ii). □

V.2.6 Complement. When $k=1$ we may do the construction above to obtain W^4 with the given properties, but $\pi_1(\partial W)$ will in general be larger than $\pi_1(W)$.

For in dimension 3 it is hard to calculate the effect of surgery on π_1.

V.2.7 Lemma. *In the construction of* (V.2.1), ∂W *is a homotopy sphere if and only if the determinant of* $M = \pm 1$.

Proof. Consider the exact sequence of $(W, \partial W)$,

$$0 \to H_q(\partial W) \xrightarrow{i_*} H_q(W) \xrightarrow{j_*} H_q(W, \partial W) \xrightarrow{\partial} H_{q-1}(\partial W) \to 0\,.$$

(We have $H_{q+1}(W, \partial W) \cong H^{q-1}(W)$ by Poincaré duality, $H_s(W)$ is zero for $s \leqq q-1$ since W is $(q-1)$-connected, so $H^{q-1}(W)=0$ by the universal coefficient formula, which produces the zero on the left.) Now by Poincaré duality the intersection pairing on $H_q(W) \otimes H_q(W, \partial W) \to \mathbb{Z}$ is non-singular, since $H_q(W)$ and $H_q(W, \partial W)$ are free, (see IV § 2, property (a) of intersection). Hence on $H_q(W) \otimes H_q(W) \to \mathbb{Z}$, the intersection product as a map $\alpha: H_q(W) \to \operatorname{Hom}(H_q(W), \mathbb{Z})$ is a monomorphism if and only if image $i_* = \ker j_* = 0$, and is onto if and only if $\partial = 0$. Since i_* is a monomorphism and ∂ is onto, it follows that α is a monomorphism if and only if $H_q(\partial W)=0$ and α is onto if and only if $H_{q-1}(\partial W)=0$. So α is an isomorphism if and only if $H_q(\partial W) = H_{q-1}(\partial W) = 0$ and ∂W is a homotopy sphere (since it is a closed $(q-2)$-connected manifold). But α is an isomorphism if and only if $\det M = \pm 1$. □

Now consider the following 8×8 matrix (see Hirzebruch [31])

$$M_0 = \begin{pmatrix} 2 & 1 & & & & & & \\ 1 & 2 & 1 & & & & & \\ & 1 & 2 & 1 & & & & \\ & & 1 & 2 & 1 & & & \\ & & & 1 & 2 & 1 & 0 & 1 \\ & & & & 1 & 2 & 1 & 0 \\ & & & & 0 & 1 & 2 & 0 \\ & & & & 1 & 0 & 0 & 2 \end{pmatrix}$$

with zeros in the blank area.

V.2.8 Lemma. *M_0 is symmetric, even on the diagonal,* $\det M_0 = 1$ *and signature* $M = 8$.

Proof. The first two statements are obvious. To prove $\det M_0 = 1$ and $\operatorname{sgn} M_0 = 8$, we will perform elementary transformations on M_0 by subtracting λ(i-th row) from j-th row, then λ(i-th column) from the j-th column, λ a rational number. This corresponds to pre- and post-multiplying M_0 by elementary matrices $I + \lambda(e_{ij})$ and $I + \lambda(e_{ji})$, where I = identity, e_{ij} is the matrix with 1 in the ij position, zero elsewhere. This process changes neither the determinant (since $\det(I + \lambda(e_{ij})) = 1$) nor the signature.

We start with the operation $\frac{1}{2}$ the first subtracted from the second. This makes the upper left corner

$$\begin{pmatrix} 2 & 0 & & \\ 0 & \frac{3}{2} & 1 & \\ & 1 & 2 & \\ & \cdots & & \end{pmatrix}$$

the remainder being unchanged. Then $\frac{2}{3}$ times second from the third gives

$$\begin{pmatrix} 2 & 0 & & & \\ 0 & \frac{3}{2} & 0 & & \\ & 0 & \frac{4}{3} & 1 & \\ & & 1 & 2 & \\ & & \cdots & & \end{pmatrix}$$

$\frac{3}{4}$ times 3rd from 4th gives

$$\begin{pmatrix} 2 & 0 & & & & \\ 0 & \frac{3}{2} & 0 & & & \\ & 0 & \frac{4}{3} & 0 & & \\ & & 0 & \frac{5}{4} & 1 & \\ & & & 1 & 2 & \\ & & & \cdots & & \end{pmatrix}$$

$\frac{4}{5}$ times 4th from 5th gives

$$\begin{pmatrix} 2 & 0 & & & & & & \\ 0 & \frac{3}{2} & 0 & & & & & \\ & 0 & \frac{4}{3} & 0 & & & & \\ & & 0 & \frac{5}{4} & 0 & & & \\ & & & 0 & \frac{6}{5} & 1 & 0 & 1 \\ & & & & 1 & 2 & 1 & 0 \\ & & & & 0 & 1 & 2 & 0 \\ & & & & 1 & 0 & 0 & 2 \end{pmatrix}.$$

Now subtract $\frac{1}{2}$ (8th) from 5th to get in the lower right corner

$$\begin{matrix} & \cdots & & & \cdots \\ & \frac{7}{10} & 1 & 0 & 0 \\ & 1 & 2 & 1 & 0 \\ & 0 & 1 & 2 & 0 \\ & 0 & 0 & 0 & 2 \\ \cdots & & & & \end{matrix}\Bigg).$$

Now carry on as before, subtract $\frac{10}{7}$ (5th) from 6th, then $\frac{7}{4}$ (6th) from 7th, with the final result, the diagonal matrix:

$$D_0 = \begin{pmatrix} 2 & & & & & & & \\ & \frac{3}{2} & & & & & & \\ & & \frac{4}{3} & & & & & \\ & & & \frac{5}{4} & & & & \\ & & & & \frac{7}{10} & & & \\ & & & & & \frac{4}{7} & & \\ & & & & & & \frac{1}{4} & \\ & & & & & & & 2 \end{pmatrix}.$$

Since D_0 is diagonal and all entries are positive, $\operatorname{sgn} D_0 = \operatorname{sgn} M_0 = 8$. One checks that the product of the entries is 1, so $\det D_0 = \det M_0 = 1$. □

Hence, putting together these results we get the Plumbing Theorem in dimensions $4k$, due to Milnor.

V.2.9 Theorem. *Let $k > 1$. There is a normal map (f, b),*

$$f:(W, \partial W) \to (D^{4k}, S^{4k-1})$$

such that $(f|\partial W)$ is a homotopy equivalence and $\sigma(f, b) = 1$.

Proof. Let $(W^{4k}, \partial W)$ be the manifold with boundary constructed in (V.2.1) using the matrix M_0. Since $\det M_0 = 1$ by (V.2.8), it follows from (V.2.7) that ∂W is a homotopy sphere. By (V.2.1) (ii), (iii) the intersection product on $K_{2k}(W)$ or $(\ ,\ )$ on $K^{2k}(W, \partial W)$ has matrix M_0, and by (V.2.8), $\operatorname{sgn} M_0 = 8$. Hence if (f, b) is the normal map

$$f:(W, \partial W) \to (D^{4k}, S^{4k-1})$$

of (V.2.1) (iii), it follows that $\sigma(f, b) = \frac{1}{8} I(f) = \frac{1}{8} \operatorname{sgn} M_0 = 1$. □

It is interesting to note that if we plumb by the matrix M_0 the result W is already $(2k-1)$-connected with ∂W a homotopy sphere. In fact we have the following graphical analysis, observed by Hirzebruch:

If we plumb together n q-disk bundles over spheres S^q we represent each sphere S^q_i by a vertex of a graph, and join the two vertices by an edge for each point of intersection. Thus M_0 is represented for example by the graph

(This is the Dynkin diagram of the exceptional Lie group E_8.)

V.2.10 Remark. The plumbed manifold has a one skeleton of the same homotopy type as the graph. In particular it is 1-connected if and only if the graph is 1-connected.

If we place on each vertex of the graph the self intersection number, or more generally, the bundle over S^q_i, then the graph describes plumbing of disk bundles over spheres completely.

To describe the plumbing necessary in dimension $4k+2$, we must pay more attention to the bundle map part of the normal map.

Let (f_i, b_i) be normal maps $f_i: M^{2q}_i \to S^{2q}$, $b_i: \nu_i \to \xi$, etc. Let $S^{2q} = D^{2q}_+ \cup D^{2q}_-$. By the homotopy extension theorem we may change f_i by a homotopy, to get f'_i such that $f'_i(x) \in D^{2q}_-$ for $x \in M_i - \operatorname{int} D^{2q}_i$, for some disk $D^{2q}_i \subset M^{2q}_i$, and $f'_i | D^{2q}_i = h_i$ is a previously given diffeomorphism of degree 1, $h_i: D^{2q}_i \to D^{2q}_+$, and we may cover the homotopy by a bundle homotopy of b_i, to a new map b'_i. If $h: D_1 \to D_2$ is the diffeomorphism defined by $h_2^{-1} h_1 = h$ then h is covered by a bundle map $c: \nu_1 | D_1 \to \nu_2 | D_2$ in a natural way. Then we may change b'_1 by a bundle homotopy to b''_1 so that $b'_2 c = b''_1$ over D^{2q}_1 (since D^{2q}_1 is contractible).

Now take the normal map (f, b), $f: S^q \times S^q \to S^{2q}$ with bundle map b coming from normal line bundles in S^{2q+1}. Obviously, (f, b) is normally cobordant to an equivalence in two different ways: by

$$(D^{q+1} \times S^q - \operatorname{int} D^{2q+1}) \quad \text{or} \quad (S^q \times D^{q+1} - \operatorname{int} D^{2q+1}).$$

Let $D_0^{2q} = D^q \times D^q$ be a neighborhood of a point $(x_0, x_0) \in S^q \times S^q$ with $D^q \times 0$ consisting of points (y, y), i.e. a product neighborhood of a point on the diagonal, $D^q \times 0$ being a neighborhood of (x_0, x_0) in the diagonal $\Delta S^q \subset S^q \times S^q$, $y \times D^q$ being the normal disks to ΔS^q. Consider D_+^{2q} as $D^q \times D^q$, where $S^{2q} = D_+^{2q} \cup D_-^{2q}$. Using the homotopy extension and bundle covering homotopy theorems as above, we may change (f, b) by a homotopy to (f_1, b_1) so that $f_1 | D_0 : D^q \times D^q \to D^q \times D^q = D_+^{2q}$ is the identity, $b_1 | v | D_0 = \text{identity}$, and $f_1(S^q \times S^q - \operatorname{int} D_0) \subset D_-^{2q}$. We may similarly change (f, b) by a homotopy to (f_2, b_2) so that

$$f_2 | D_0 : D^q \times D^q \to D^q \times D^q = D_+^{2q}$$

is I_+, b_2 is the bundle map induced by I_+ on $v | D_0^{2q}$, and

$$f_2(S^q \times S^q - \operatorname{int} D_0^{2q}) \subset D_-^{2q}.$$

Let $D_1^{2q} \subset S^q \times S^q - \operatorname{int} D_0^{2q}$ be a disk disjoint from D_0. Then the restrictions define normal maps

$$(f_i', b_i') : (S^q \times S^q - \operatorname{int} D_1, S_1^{2q-1}) \to (D_+^{2q}, S_+^{2q-1}).$$

Let E be a tubular neighborhood of ΔS^q in $S^q \times S^q$, with $D_0^{2q} = \pi^{-1}(D^q)$, $D^q \subset \Delta S^q$, $\pi : E \to \Delta S^q$ the projection. We may assume that

$$f_i'(S^q \times S^q - E) \subset S_+^{2q-1}.$$

Now if we identify in two copies E_1, E_2 by the diffeomorphism I_+ on $D_{01} \subset E_1$ with $D_{02} \subset E_2$, i.e. plumb E_1 and E_2 together, to get $U = E_1 \square E_2$, then the restrictions of (f_1, b_1) and (f_2, b_2) agree on $E_1 \cap E_2 = D_0^{2q}$, so that the union defines a normal map

$$(g, c),\ g : (U, \partial U) \to (D^{2q}, S^{2q-1}).$$

Now we have the Plumbing Theorem for dimensions $4k+2$, due to Kervaire [35].

V.2.11 Theorem. *For q odd, ∂U^{2q} is a homotopy sphere, and $\sigma(g, c) = 1$.*

Proof. The normal bundle of the diagonal $\Delta S^q \subset S^q \times S^q$ is equivalent to the tangent bundle τ_{S^q} of S^q. If q is odd, then $\chi(\tau_{S^q}) = 0$, so the intersection matrix of U is $\begin{pmatrix} 0 & 1 \\ -1 & 0 \end{pmatrix}$ (skew symmetric since q is odd). It follows easily from van Kampen's theorem that for $q > 1$, U is 1-connected, since $E_1 \cap E_2 = D_0^{2q}$ and E_1 and E_2 are 1-connected if $q > 1$. A similar argument shows ∂U is 1-connected for $q > 1$. If $q = 1$, ∂U is a closed and connected 1-manifold hence a circle. If $q > 1$, then ∂U is 1-connected and the intersection matrix of U has determinant $+1$, so ∂U is a homotopy sphere by (V.2.7). (Note that for q even, the intersection matrix becomes $\begin{pmatrix} 2 & 1 \\ 1 & 2 \end{pmatrix}$, so it has determinant $= 3$.)

Now we would like to compute the quadratic form ψ defined in III § 4. Since $E_i \subset U$, there is a natural collapsing map $\eta_i: U/\partial U \to E_i/\partial E_i$, and $[U] \cap \eta_i^*(U_i) = j_{i*}[S_i^q]$ by (IV.2.8) (where j_i denotes inclusion). Since $j_{i*}[S_i^q], i=1,2$ are a basis for $H_q(U), j_{1*}[S_1^q] \cdot j_{2*}[S_2^q] = 1, j_{i*}[S_i^q] \cdot j_{i*}[S_i^q] = 0$, $i=1,2$, it follows that $x_i = \eta_i^*(U_i)$, $i=1,2$ is a symplectic basis for $H^q(U, \partial U; \mathbb{Z}_2)$, so $\sigma(g,c) = c(g,c) = c(\psi) = \psi(x_1)\,\psi(x_2)$.

We need the following:

V.2.12 Lemma. *Let $V^m \subset W^m$ be the inclusion of a submanifold of the same dimension, V and W manifolds with boundary and let $\eta: W/\partial W \to V/\partial V$ be the natural collapsing man. Let ν = normal bundle of W in D^{m+k}, so $\nu | V$ is the normal bundle of V in D^{m+k}. Then the inclusion $T(\nu | V) \to T(\nu)$ is Spanier-Whitehead S-dual to $\eta: W/\partial W \to V/\partial V$.*

Proof. Embed $(W, \partial W) \subset (D^{m+k}, S^{m+k-1})$ in such a way that

$$(V, \partial V) \subset (D_1^{m+k}, S_1^{m+k-1})$$

where $D_1^{m+k} \subset D^{m+k}$ is a disk of $\frac{1}{2}$ the radius. Then $D^{m+k} - W$ is the complement of $W \cup D_0^{m+k}$ in S^{m+k} where $D \cup D_0 = S^{m+k}$, $D \cap D_0 = S^{m+k-1}$, so $D_0 \cap W = \partial W$ and $W \cup D_0^{m+k}$ is homotopy equivalent to $W/\partial W$. Hence $D^{m+k} - W$ is S-dual to $W/\partial W$. Similarly, $D_1 - V$ is S-dual to $V/\partial V$, and the inclusion $D_1 - V$ into $D - W$ is S-dual to the inclusion $W \cup D_0$ into $W \cup (\overline{D - D_1}) \cup D_0$. But the latter inclusion is homotopy equivalent to the collapsing map $\eta: W/\partial W \to V/\partial V$.

Now the inclusion $D_1 - V \subset D - W$ and the inclusion $E(\nu | V) \subset E(\nu)$ coincide with the inclusion $D_1 \subset D$. Then $j: D_1/\overline{D_1 - E(\nu | V)} \to D/\overline{D - E(\nu)}$ is the suspension of the inclusion $D_1 - E(\nu | V) \subset D - E(\nu)$, and j is also the inclusion

$$T(\nu | V) = E(\nu | V)/E_0(\nu | V) \to T(\nu) = E(\nu)/E_0(\nu)\,.$$

Hence the inclusion $T(\nu | V) \subset T(\nu)$ is S-dual (in S^{m+k+1}) to

$$\eta: W/\partial W \to V/\partial V. \quad \square$$

V.2.13 Lemma. *Let $V^{2q} \subset W^{2q}$ be a submanifold with boundary, and let (f,b) be a normal map, $f: (W, \partial W) \to (A, B)$ such that $f(W - V) \subset B$ so that $(f | V, b | V)$, $f | V: (V, \partial V) \to (A, B)$ is also a normal map. Let $\eta: W/\partial W \to V/\partial V$ be the collapsing map, and let*

$$\psi: K^q(W, \partial W) \to \mathbb{Z}_2, \qquad \psi': K^q(V, \partial V) \to \mathbb{Z}_2$$

be the quadratic forms of III § 4. *Then $\psi(\eta^*(x)) = \psi'(x)$ for $x \in K^q(V, \partial V)$.*

Proof. Recall from III § 4 that $T(b): T(\nu) \to T(\xi)$, (ξ over A) is S-dual to a map $g: \Sigma^k(A/B) \to \Sigma^k(W/\partial W)$, and that $K^q(W, \partial W) \cong \ker(g^* \Sigma^k)^q$ (see III 4.1). Since $j: T(\nu|V) \to T(\nu)$ is S-dual to $\eta: W/\partial W \to V/\partial V$ by (V.2.12), it follows that $T(b)j: T(\nu|V) \to T(\xi)$ is S-dual to

$$(\Sigma^k \eta) g: \Sigma^k(A/B) \to \Sigma^k(V/\partial V).$$

It follows that $\eta^*(K^q(V, \partial V)) \subset K^q(W, \partial W)$. Hence if $x \in K^q(V, \partial V)$, ψ is defined on $\eta^*(x)$. Let $\varphi: V/\partial V \to K(\mathbb{Z}_2, q)$ be such that $\varphi^*(\iota) = x$. Then $\psi'(x) = (\mathrm{Sq}_h^{q+1}(\Sigma^k(\iota)))(\Sigma^k[A])$, (see III § 4) where $h = (\Sigma^k \varphi)(\Sigma^k \eta) g$. Also since $(\Sigma^k \varphi)(\Sigma^k \eta) = \Sigma^k(\varphi\eta)$, and $(\varphi\eta)^*(\iota) = \eta^*(\varphi^* \iota) = \eta^* x$, it follows that the same formula defines $\psi(\eta^*(x))$, so $\psi(\eta^* x) = \psi'(x)$. □

Now we return to the proof of (V.2.11), and we show that

$$\psi(x_1) = \psi(x_2) = 1,$$

so that $\sigma(g, c) = 1$.

From (V.2.13), we deduce that $\psi(x_i) = \psi'(U_i)$, where $U_i \in H^q(E_i, \partial E_i)$ is the Thom class, ψ' is the quadratic form associated to the normal map $(g|E_i, c|E_i)$. By construction $g|E_i = f_i'|E_i$, where

$$f_i': (S^q \times S^q - \operatorname{int} D_1^{2q}, S_1^{2q-1}) \to (D^{2q}, S^{2q-1}),$$

and (f_i', b_i') is homotopic to (and hence normally cobordant to) (f', b'), which is the restriction of (f, b), $f: S^q \times S^q \to S^{2q}$, described above. From the construction of (f, b) and the two different normal cobordisms of (f, b) to an equivalence, we may deduce that if $y \otimes 1, 1 \otimes y \in H^q(S^q \times S^q)$, then $\psi''(y \otimes 1) = \psi''(1 \otimes y) = 0$, ψ'' being the quadratic form associated to (f, b). Hence $\psi''(y \otimes 1 + 1 \otimes y) = 1$. But if U = Thom class of the normal bundle of the diagonal $\Delta S^q \subset S^q \times S^q$, $U \in H^q(E/\partial E)$, $\eta: S^q \times S^q \to E/\partial E$, then $\eta^* U = y \otimes 1 + 1 \otimes y$. It is clear that ψ'' defined by (f, b), $f: S^q \times S^q \to S^{2q}$ is the same as ψ''' defined by (f', b'), the restriction of (f, b),

$$f': (S^q \times S^q - \operatorname{int} D_1^{2q}, S_1^{2q-1}) \to (D^{2q}, S^{2q-1}),$$

so that by (V.2.13), $\psi'''(y \otimes 1 + 1 \otimes y) = \psi'(U) = 1$. Hence $\psi(x_i) = \psi'(U_i) = 1$, $i = 1, 2$, $\sigma(g, c) = \psi(x_1)\psi(x_2) = 1$, and (V.2.11) is proved. □

Bibliography

1. Abraham, R., Robbin, J.: Transversal mappings and flows. New York: W. A. Benjamin 1967.
2. Arf, C.: Untersuchungen über quadratische Formen in Körpern der Charakteristik 2. Crelles Math. J. **183**, 148–167 (1941).
3. Artin, E.: Geometric algebra. New York: Interscience 1957.
4. Atiyah: Thom complexes. Proc. London Math. Soc. **11**, 291–310 (1961).
5. Bott, R., Milnor, J.: On the parallelizability of the spheres. Bull. Amer. Math. Soc. **64**, 87–89 (1958).
6. Browder, W.: Homotopy type of differentiable manifolds. Proceedings of the Aarhus Symposium, 1962, 42–46.
7. — The Kervaire invariant of framed manifolds and its generalization. Ann. Math. **90**, 157–186 (1969).
8. — Embedding 1-connected manifolds. Bull. Amer. Math. Soc. **72**, 225–231 (1966).
9. — Embedding smooth manifolds. Proceedings of the International Congress of Mathematicians, Moscow, 1966.
10. — Manifolds with $\pi_1 = Z$. Bull. Amer. Math. Soc. **72**, 238–244 (1966).
11. — Diffeomorphisms of 1-connected manifolds. Trans. Amer. Math. Soc. **128**, 155–163. (1967).
12. — Surgery and the theory of differentiable transformation groups. Proceedings of the Tulane Symposium on Transformation Groups, 1967, pp. 1–46. Berlin-Heidelberg-New York: Springer 1968.
13. — Hirsch, M.: Surgery on p.l. manifolds and applications. Bull. Amer. Math. Soc. **72**, 959–964 (1966).
14. Brown, E. H.: Cohomology theories. Ann. Math. **75**, 467–484 (1962).
15. — Nonexistence of low dimension relations between Stiefel-Whitney classes. Trans. Amer. Math. Soc. **104**, 374–382 (1962).
16. Cartan, H.: Seminare H. Cartan 1954–1955, Algèbres d'Eilenberg-MacLane et homotopie. Paris: École Normale Supérieure 1956.
17. Cerf, J.: Topologie de certains espaces de plongements. Bull. Soc. Math. France **89**, 227–380 (1961).
18. Conner, P., Floyd, E. E.: Differentiable periodic maps. Berlin-Göttingen-Heidelberg-New York: Springer 1964.
19. Dold, A.: Über fasernweise Homotopie äquivalenz von Faserräumen. Math. Z. **62**, 111–136 (1955).
20. — Relations between ordinary and extraordinary homology. Proc. Aarhus Symposium 1962, 2–9.
21. Eilenberg, S., Silber, J.: On products of complexes. Amer. J. Math. **75**, 200–204 (1953).
22. — Steenrod, N.: Foundations of algebraic topology. Princeton: University Press 1952.
23. Golo, V.: Smooth structures on manifolds with boundary. Dokl. Akad. Nauk. S.S.S.R., **157**, 22–25 (1964) = Soviet Math. Doklady, **5**, 862–866 (1964).

24. Haefliger, A.: Knotted $(4k-1)$ – spheres in $6k$ – space. Ann. Math. **75**, 452–466 (1962).
25. — Differentiable embeddings of S^n in S^{n+q} for $q>2$. Ann. Math. **83**, 402–436 (1966).
26. — Knotted spheres and related geometric problems. Proc. of Int. Congress of Mathematicians, Moscow 1966.
27. — Enlacements de sphères en codimension supérieure à 2. Comment. Math. Helv. **41**, 51–72 (1966).
28. Hilton, P.J.: Homotopy theory and duality. New York: Gordon and Breach 1966.
29. Hirsch, M., Mazur, B.: Smoothing theory, mimeographed notes. Cambridge University 1964.
30. Hirzebruch, F.: New topological methods in algebraic geometry. 3rd Ed. Berlin-Heidelberg-New York: Springer 1966.
31. — Differentiable manifolds and quadratic forms, Lecture notes. New York: Marcel Decker 1962.
32. Husemoller, D.: Fibre bundles. New York: McGraw Hill 1966.
33. James, I.M.: On H-spaces and their homotopy groups. Quart. J. Math. **11**, 161–179 (1960).
34. Kervaire, M., Milnor, J.: Groups of homotopy spheres I. Ann. Math. **77**, 504–537 (1963).
35. — A manifold which does not admit any differentiable structure. Comment. Math. Helv. **34**, 257–270 (1960).
36. — Non-parallelizability of the n-sphere for $n>7$. Proc. Nat. Acad. Sci. U.S.A. **44**, 280–283 (1958).
37. — Milnor, J.: Bernoulli numbers, homotopy groups and a theorem of Rohlin. Proc. Int. Congress of Math., Edinborough, 1958.
38. Levine, J.: On differentiable embeddings of simply connected manifolds. Bull. Amer. Math. Soc. **69**, 806–809 (1963).
39. — A classification of differentiable knots. Ann. Math. **82**, 15–50 (1965).
40. Mazur, B.: Differential topology from the point of view of simple homotopy theory. Inst. Hautes Études Sci. Publ. Math. **15** (1966).
41. Milnor, J.: Morse theory. Annals of Math. Studies No. 51. Princeton: University Press 1963.
42. — Lectures on the h-cobordism theorem, notes by L. Siebenmann and J. Sondow. Princeton: University Press 1965.
43. — Spanier, E.: Two remarks on fibre homotopy type. Pacific J. Math. **10**, 585–590 (1960).
44. — Characteristic classes, mimeographed notes. Princeton University 1957.
45. — A procedure for killing the homotopy groups of differentiable manifolds. Symposia in Pure Math., Amer. Math. Soc. **3**, 39–55 (1961).
46. — On simply connected 4-manifolds. Symp. Topologia Algebraica, Mexico 1958, 122–128.
47. — Differentiable manifolds which are homotopy spheres, mimeographed notes. Princeton 1958.
48. — Microbundles. Topology **3**, 53–80 (1964).
49. Novikov, S. P.: Diffeomorphisms of simply connected manifolds. Soviet Math. Dokl. **3**, 540–543 (1962), = Dokl. Akad. Nauk. S.S.S.R. **143**, 1046–1049 (1962).
50. — Homotopy equivalent smooth manifolds I. AMS Translations **48**, 271–396 (1965). = Izv. Akad. Nauk. S.S.S.R. Serv. Mat. **28**, 365–474 (1964).
51. Palais, R.: Extending diffeomorphisms. Proc. Amer. Math. Soc. **11**, 274–277 (1960).
52. Serre, J. P.: Cohomologie modulo 2 des complexes d'Eilenberg-MacLane. Comment. Math. Helv. **27**, 198–231 (1953).
53. — Groupes d'homotopie et classes de groupes abéliens. Ann. Math. **58**, 258–294 (1953).

54. Smale, S.: On the structure of manifolds. Amer. J. Math. **84**, 387–399 (1962).
55. Spanier, E. H.: Algebraic topology. New York: McGraw Hill 1966.
56. — Function spaces and duality. Ann. Math. **70**, 338–378 (1959).
57. Spivak, M.: Spaces satisfying Poincaré duality. Topology **6**, 77–104 (1967).
58. Stasheff, J.: A classification theorem for fibre spaces. Topology **2**, 239–246 (1963).
59. Steenrod, N., Epstein, D. B. A.: Cohomology operations. Annals of Math. Studies No. 50, Princeton Univ. Press 1962.
60. — The topology of fibre bundles. Princeton Math. Series 14, Princeton: University Press 1951.
61. — Cohomology invariants of mappings. Ann. Math. **50**, 954–988 (1949).
62. Sullivan, D.: Triangulating homotopy equivalences. Thesis Princeton University 1965.
63. — On the Hauptvermutung for manifolds. Bull. Amer. Math. Soc. **73**, 598–600 (1967).
64. Thom, R.: Quelques propriétés globales des variétés differentiables. Comment. Math. Helv. **28**, 17–86 (1954).
65. Wall, C. T. C.: An extension of a result of Novikov and Browder. Amer. J. Math. **88**, 20–32 (1966).
66. — Surgery on non-simply connected manifolds. Ann. Math. **84**, 217–276 (1966), and surgery on compact manifolds. London: Academic Press 1970.
67. — Poincaré complexes I. Ann. Math. **86**, 213–245 (1967).
68. Wallace, A. H.: Modifications and cobounding manifolds. Canad. J. Math. **12**, 503–528 (1960).

Subject Index

Ergebnisse der Mathematik und ihrer Grenzgebiete

1. Bachmann: Transfinite Zahlen.
2. Miranda: Partial Differential Equations of Elliptic Type.
4. Samuel: Méthodes d'Algèbre Abstraite en Géométrie Algébrique.
5. Dieudonné: La Géométrie des Groupes Classiques.
6. Roth: Algebraic Threefolds with Special Regard to Problems of Rationality.
7. Ostmann: Additive Zahlentheorie. 1. Teil: Allgemeine Untersuchungen.
8. Wittich: Neuere Untersuchungen über eindeutige analytische Funktionen.
11. Ostmann: Additive Zahlentheorie. 2. Teil: Spezielle Zahlenmengen.
13. Segre: Some Properties of Differentiable Varieties and Transformations.
14. Coxeter/Moser: Generators and Relations for Discrete Groups.
15. Zeller/Beckmann: Theorie der Limitierungsverfahren.
16. Cesari: Asymptotic Behavior and Stability Problems in Ordinary Differential Equations.
17. Severi: Il teorema di Riemann-Roch per curve-superficie e varietà questioni collegate.
18. Jenkins: Univalent Functions and Conformal Mapping.
19. Boas/Buck: Polynomial Expansions of Analytic Functions.
20. Bruck: A Survey of Binary Systems.
21. Day: Normed Linear Spaces.
23. Bergmann: Integral Operators in the Theory of Linear Partial Differential Equations.
25. Sikorski: Boolean Algebras.
26. Künzi: Quasikonforme Abbildungen.
27. Schatten: Norm Ideals of Completely Continuous Operators.
28. Noshiro: Cluster Sets.
30. Beckenbach/Bellman: Inequalities.
31. Wolfowitz: Coding Theorems of Information Theory.
32. Constantinescu/Cornea: Ideale Ränder Riemannscher Flächen.
33. Conner/Floyd: Differentiable Periodic Maps.
34. Mumford: Geometric Invariant Theory.
35. Gabriel/Zisman: Calculus of Fractions and Homotopy Theory.
36. Putnam: Commutation Properties of Hilbert Space Operators and Related Topics.
37. Neumann: Varieties of Groups.
38. Boas: Integrability Theorems for Trigonometric Transforms.
39. Sz.-Nagy: Spektraldarstellung linearer Transformationen des Hilbertschen Raumes.
40. Seligman: Modular Lie Algebras.
41. Deuring: Algebren.
42. Schütte: Vollständige Systeme modaler und intuitionistischer Logik.
43. Smullyan: First-Order Logic.
44. Dembowski: Finite Geometries.
45. Linnik: Ergodic Properties of Algebraic Fields.
46. Krull: Idealtheorie.
47. Nachbin: Topology on Spaces of Holomorphic Mappings.
48. A. Ionescu Tulcea/C. Ionescu Tulcea: Topics in the Theory of Lifting.
49. Hayes/Pauc: Derivation and Martingales.
50. Kahane: Séries de Fourier Absolument Convergentes.
51. Behnke/Thullen: Theorie der Funktionen mehrerer komplexer Veränderlichen.
52. Wilf: Finite Sections of Some Classical Inequalities.
53. Ramis: Sous-ensembles analytiques d'une variété banachique complexe.
54. Busemann: Recent Synthetic Differential Geometry.

55. Walter: Differential and Integral Inequalities.
56. Monna: Analyse non-archimédienne.
57. Alfsen: Compact Convex Sets and Boundary Integrals.
58. Greco/Salmon: Topics in $\mathfrak{m}$-Adic Topologies.
59. López de Medrano: Involutions on Manifolds.
60. Sakai: C*-Algebras and W*-Algebras.
61. Zariski: Algebraic Surfaces.
62. Robinson: Finiteness Conditions and Generalized Soluble Groups, Part 1.
63. Robinson: Finiteness Conditions and Generalized Soluble Groups, Part 2.
64. Hakim: Topos annelés et schémas relatifs.
65. Browder: Surgery on Simply-Connected Manifolds.
66. Pietsch: Nuclear Locally Convex Spaces.
67. Dellacherie: Capacités et processus stochastiques.
68. Raghunathan: Discrete Subgroups of Lie Groups.
69. Rourke/Sanderson: Introduction to Piecewise-Linear Topology.
70. Kobayashi: Transformation Groups in Differential Geometry.